AUTOMATED TEST METHODS FOR FRACTURE AND FATIGUE CRACK GROWTH

A symposium
sponsored by
ASTM Committees E-9 on
Fatigue and E-24 on
Fracture Testing
Pittsburgh, PA, 7–8 Nov. 1983

ASTM SPECIAL TECHNICAL PUBLICATION 877
W. H. Cullen, Materials Engineering Associates,
R. W. Landgraf, Southfield, Mich.,
L. R. Kaisand, General Electric R&D Center, and
J. H. Underwood, Benet Weapons Laboratory,
editors

ASTM Publication Code Number (PCN)
04-877000-30

 1916 Race Street, Philadelphia, PA 19103

Library of Congress Cataloging in Publication Data

Automated test methods for fracture and fatigue crack growth.

(ASTM special technical publication; 877)
"ASTM publication code number (PCN) 04-877000-30."
Includes bibliographies and index.
1. Materials—Fatigue—Congresses.   2. Fracture
mechanics—Congresses.   I. Cullen, W. H.   II. Landgraf, R. W.,   III. Kaisand,
L. R.,   IV. Underwood, J. H.,   V. American Society for Testing and Materials.
Committee E-9 on Fatigue.   VI. ASTM Committee E-24 on Fracture Testing.   VII.
Series.
TA418.38.A98   1985   620.1′123   85-15710
ISBN 0-8031-0421-9

NOTE
The Society is not responsible, as a body,
for the statements and opinions
advanced in this publication.

Printed in Ann Arbor, MI
October 1985

Foreword

The symposium on Automated Test Methods for Fracture and Fatigue Crack Growth was held in Pittsburgh, Pennsylvania, 7–8 November 1983. ASTM Committees E-9 on Fatigue and E-24 on Fracture Testing sponsored the symposium. W. H. Cullen, Materials Engineering Associates, R. W. Landgraf, Southfield, Michigan, L. R. Kaisand, General Electric R&D Center, and J. H. Underwood, Benet Weapons Laboratory, presided as symposium chairmen and are editors of this publication.

Related
ASTM Publications

Methods and Models for Predicting Fatigue Crack Growth Under Random
Loading, STP 748 (1981), 04-748000-30

Fatigue Crack Growth Measurement and Data Analysis, STP 738 (1981),
04-738000-30

Effect of Load Variables on Fatigue Crack Initiation and Propagation,
STP 714 (1980), 04-714000-30

Part-Through Crack Fatigue Life Prediction, STP 687 (1979), 04-687000-30

Flaw Growth and Fracture (10th Conference), STP 631 (1977), 04-631000-30

Fatigue Crack Growth Under Spectrum Loads, STP 595 (1976), 04-595000-30

Mechanics of Crack Growth, STP 590 (1976), 04-590000-30

Fracture Touchness and Slow-Stable Cracking (8th Conference), STP 559
(1974), 04-559000-30

Stress Analysis and Growth of Cracks, STP 513 (1973), 04-513000-30

A Note of Appreciation
to Reviewers

The quality of the papers that appear in this publication reflects not only the obvious efforts of the authors but also the unheralded, though essential, work of the reviewers. On behalf of ASTM we acknowledge with appreciation their dedication to high professional standards and their sacrifice of time and effort.

ASTM Committee on Publications

ASTM Editorial Staff

Helen M. Hoersch
Janet R. Schroeder
Kathleen A. Greene
Bill Benzing

Contents

Overview

With the rapid advances in the incorporation of automated data acquisition and processing capabilities into many mechanical testing laboratories, it has become increasingly possible to conduct many experiments entirely under computer control. Computers, data loggers, and measurement and control processors, together with load cells, displacement gages, and their conditioning circuits, or electric potential-drop systems, have created an entirely new set of opportunities for the improvement of fatigue and fracture tests that were formerly conducted under essentially manual control using optical or simple analog methods of data acquisition. The existing ASTM standards for fatigue and fracture testing, while they are carefully worded so as to allow incorporation of automated techniques, do not specifically set down the methods for performing tests with fully automated test facilities. Since automated testing is possibly the present, or certainly the future, for many laboratories, many of the applicable standards face rewriting, or will require annexes (appendices) to specifically establish the requirements for automated methodologies.

The Symposium on Automated Test Methods for Fracture and Fatigue Crack Growth was held in Pittsburgh, PA on 7–8 November 1983 to provide a forum for researchers using automated systems to describe their techniques, and to discuss especially the methods used to establish conformance to, or exceed the requirements of, the various ASTM standards for fatigue and fracture which were used as the basis for the test. The contributors were asked to provide descriptions of the techniques used in their test systems, and to address how they qualified their systems to assure that the data conformed to the existing ASTM standard test practices. The contributions to the symposium covered a wide range of techniques and test objectives, and were provided by scientists from laboratories all over the world. The symposium was very well attended at all three sessions. The first two sessions addressed techniques used for fatigue and fatigue crack growth rate testing, and the final session dealt with techniques for fracture testing.

The arrangement of contributions to this STP follows the order of presentation at the symposium. In the final analysis, the authors provided more description of their test systems, and somewhat less description of the ways in which the systems conformed to, or exceeded, the requirements of the applicable ASTM standards. Thus, techniques for assuring accuracy and precision of these automated methods have still not been subjected to the kind of

1

open forum which may be required before there is general acceptance of a particular methodology.

Systems for Fatigue and Fatigue Crack Growth Testing

Thirteen papers have been contributed in this category.

The paper by Miller and co-authors from the University of Illinois takes advantage of this university's long involvement in the development of computerized test control and data acquisition instrumentation. The history of laboratory computers is reviewed, and a description of a current system design is provided. Several computer-to-computer communication protocols are mentioned, since these are necessary for passing data from one laboratory location to another. Lastly, the general impact of these current advances on standards writing is discussed.

The use of a personal computer to monitor sustained load cracking test progress at several test stands is described by Meyn et al. This system has the advantage that data are acquired in proportion to the rate of change of the test specimen response; that is, when the loads or displacements of the specimen are changing rapidly, data acquisition is frequent, but when crack extension in the specimen is in an incubation stage, data acquisition is quite infrequent. The criteria for rejection of false data are discussed.

Vecchio and colleagues describe an automated system for fatigue crack growth that has been used on compact and three-point bend specimens over a wide range of growth rates, for both metals and polymers. The influence of overloads on crack closure, and therefore on the compliance technique for monitoring crack extension, is discussed.

Catlin and co-workers discuss a novel approach to the use of direct-current potential-drop methods in aqueous environments. Careful consideration has been given to the possibility that the currents and voltage levels used to provide the potential-drop capability might interfere with the corrosion potential of the specimen. This paper also describes the techniques used to assure that the systems have long-term stability, low noise, and can be applied to a number of specimen geometries and crack shapes.

Scientists at larger laboratories may be interested in the discussion of a distributed system approach to computerized test practice described in a contribution by Topp and Dover. In particular, the authors discuss their application of an alternating-current method of crack extension determination, and its application to somewhat large test specimens and nonstandard geometries, such as tubular joints, and threaded sections.

One of the most tedious of the fatigue crack growth experiments is the determination of near-threshold data. Systems for this application are described in contributions by Sooley and Hoeppner and by McGowan and Keating. The McGowan/Keating system measures crack extension by both the compliance and potential-drop methods, and controls the rate of change of applied cyclic stress-intensity factor, ΔK, to a user-selected value. The procedures for selecting the locations for the potential probes, and the methods for assuring that the crack is fully open, before making a potential measurement, are pointed out. Sooley and Hoeppner discuss their approach to the near-threshold growth rate test practice using a very inexpensive controller. The authors indicate that this system meets the existing requirements of the ASTM Test Method for Constant-Load-Amplitude Fatigue Crack Growth Rates Above 10^{-8} m/Cycle (E647-83), and the proposed requirements for threshold testing.

Fatigue crack initiation from a blunt notch is a study requiring extremely high sensitivity measurement techniques, and a paper by Kondo and Endo presents a unique approach to this problem. Compact specimens were instrumented with back-face strain gages, and a special analog processing circuit was constructed to subtract an offset voltage from the resultant signal output, thus allowing high amplification of the incremental output from the gage. Using this system, the authors were able to detect extremely small crack extensions, and found that initiation from a blunt notch occurred much earlier in the specimen life than had been expected.

There are some attractive advantages to conducting a constant ΔK experiment, making it easier to concentrate on the other critical variables that may affect fatigue crack growth rates. Van Der Sluys and Futato review their experiences with a four-station data acquisition system that controls all the aspects of test practice, from setup through test termination, including changes in test frequency and loading parameters that may be required at various intervals in the test schedule.

Fatigue crack growth of part-through cracks in flat specimens, sometimes called surface-defected panels, is very applicable in the sense that these flaws are more geometrically similar to those that actually occur in service. Van-Stone and Richardson describe very carefully the experimental methods and calculations which are involved in the testing of such specimens, and discuss some of the techniques needed to derive the crack aspect ratio. They also discuss the effect of net section stress on aspect ratio and growth rates.

The use of surface-bonded resistance gages to measure crack extension is described in a contribution by Liaw and co-workers. Various forms of these gages have been used in air, salt water, and wet hydrogen, and a plasma-sprayed version is being evaluated for high-temperature testing. The gages have been used to monitor growth of short cracks, and have also been shown to generate, for longer cracks, data that are in good agreement with data from compliance and optical methods of crack length determination.

Testing of irradiated materials is the subject of a paper by Tjoa and co-authors. Of necessity, these specimens must be tested remotely, and use of both compliance and potential-drop methods are described. The discussion focusses on the computer algorithm used and on the errors which may be incurred in either method.

Cheng and Read discuss a system for high-frequency constant-amplitude and near-threshold testing that has been used for testing cast stainless steels at liquid helium temperatures. This system utilizes a digitizing oscilloscope to capture the rapidly varying load and displacement signals. The use of an effective modulus to match the computer calculated and optically measured crack lengths is discussed, along with the requirements for overprogramming the servohydraulic system to achieve the high test frequencies.

Systems for Fracture Testing

Four papers have been contributed in this category.

The first paper provides an interesting crossover since it discusses the elastic plastic parameter, J-integral, as it can be used in low-cycle fatigue. Joyce and Sutton describe the automated test method used to calculate and apply the desired J-integral range, and to measure and correct the loads for crack closure, in real time.

Jolles describes an automated system for R-curve measurement using either compact or bend specimens. The criteria for hardware selection based on the required sensitivity are discussed, and the use of the direct-current electric potential-drop method is presented. The potential-drop technique eliminates the need for frequent partial unloadings in order to obtain a compliance measurement.

Saario and co-workers present results on the elastic-plastic fracture testing of compact, round compact, and three-point bend specimens. An automated system has been used to carry out these tests in accordance with the proposed ASTM R-curve test procedures. The rate of load application has been shown to affect the correlation coefficient of the unloading compliance.

The final paper in this section presents a methodology for measuring the errors involved in automated systems used for fracture testing. Jablonski shows how the various contributions to errors in the load, crack opening displacement, and specimen modulus enter into the J-integral and crack extension calculations. A comparison of the results from compact and three-point bend specimens shows that the tearing modulus is different in the two geometries. The effect of side grooves and a/W ratio on the R-curve is also described in some detail.

The overall evaluation of this symposium is that there were a number of contributions which described interesting and unique approaches to the topics of automated testing, and indirect measurement of fatigue and slow-stable crack growth. However, it is obvious that there is no consensus about the exact procedures, calibration methods, or post-test data processing that would be necessary before standards can be drafted for the test methods involved in this research. However, the editors are certain that standardized test methods are feasible at this time, and in fact, at the time this overview was drafted, an effort to write an appendix for ASTM Method E 647 to incorporate compliance methods of crack length determination was underway. On the fracture side, a full-fledged standards writing effort for J-R curve determination, using the unloading compliance method, is nearing completion. It seems likely that, as time goes on, other standards for mechanical test practice will be modified or created to take advantage of computerized laboratory techniques.

W. H. Cullen

Materials Engineering Associates, Lanham, MD 20706; symposium cochairman and coeditor.

Systems for Fatigue and Fatigue Crack Growth Testing

Norman R. Miller, [1] *Dennis F. Dittmer,* [1] *and Darrell F. Socie* [1]

New Developments in Automated Materials Testing Systems

REFERENCE: Miller, N. R., Dittmer, D. F., and Socie, D. F., **"New Developments in Automated Materials Testing Systems,"** *Automated Test Methods for Fracture and Fatigue Crack Growth, ASTM STP 877,* W. H. Cullen, R. W. Landgraf, L. R. Kaisand, and J. H. Underwood, Eds., American Society for Testing and Materials, Philadelphia, 1985, pp. 9–26.

ABSTRACT: This paper traces the development of automated materials testing systems over the past ten years. The rapid reduction in computing hardware costs in recent years, coupled with fundamental improvements in computing systems design, has led to the development of a new generation of test control systems. The paper focuses on recent developments in this area at the University of Illinois at Urbana-Champaign.

The paper describes in detail a microcomputer-based controller designed to be used with a standard servohydraulic test frame. The controller uses menu-driven software which permits the operator to set up and execute tests, sample and store data, and transfer the collected data to a host computer system for data reduction and archival storage. Currently, software exists to perform standard low-cycle fatigue tests and other related test procedures. The software is designed to permit ease of operation and reduce the chance of operator error. In addition, numerous checks are performed during the course of a test to assure that the test is carried out in accord with ASTM specifications where applicable.

The paper contains a discussion of digital communications as they relate to the materials testing laboratory. The growing array of computing hardware in the testing laboratory necessitates the careful selection of communication techniques to match the needs of each application and the laboratory as a whole. The paper concludes with a discussion of the impact of new testing techniques on testing standards.

KEY WORDS: automated testing, computer control, data acquisition, data transmission, standards

The use of computers in the fatigue and fracture test laboratories has evolved steadily over the past ten years. Prior to this time, essentially no computer capabilities existed in conjunction with actual machine control or real-

[1] Associate profesor, research associate, and associate professor, respectively, Department of Mechanical and Industrial Engineering, University of Illinois at Urbana-Champaign, Urbana IL 61801.

time data acquisition tasks. The servohydraulic test frames used to conduct fatigue and fracture tests were instrumented with analog function generators, mechanical relay counters, digital volt meters, X-Y recorders, etc. In the hands of well-trained technicians, this test instrumentation sufficed to allow the execution of a considerable variety of still relevant materials tests: low-cycle fatigue, constant-amplitude crack growth, tension tests, K_{Ic} tests, etc.

This type of test instrumentation limited test control capabilities to a very narrow range of options. First, only the feedback transducer variable could be directly controlled, that is, either load, strain, or stroke; secondly, the command history was essentially limited to either a monotonic ramp or a constant-amplitude sinusoidal or triangular waveform. Also, the data acquisition instrumentation (of which the X-Y recorders were probably the most relied upon) left much to be desired. For analysis of test results it was necessary to "digitize" X-Y recorder traces by manual techniques; that is, technicians were required to pick data points off the graph paper and laboriously generate relatively small data banks of test results. Not only was this undesirable because of the consumption of time in generating the data, but also because the accuracy of the data was always uncertain due to operator subjectivity in the "data acquisition" process. The data so acquired were usually transfered to a mainframe computer system (if available) for data analysis purposes; this was accomplished by technicians inputting the data via a remote terminal or a keypunch machine. Thus, in the late 1960's, the computer served a very limited role in materials testing.

The initial two areas where materials researchers sought improved test capabilities were those of command history generation and data acquisition. Some attempts were made with analog computer systems (operational amplifier technology) to provide increased capabilities [1]. However, the results of such endeavors, although quite interesting in some cases, did not justify the time consumed in developing and setting up analog computer control or "data acquisition" systems. Such systems did not really get to the root of the problem of improving control and monitoring capabilities.

One of the first computer systems to successfully address this problem was partially developed at the University of Illinois by MTS Systems Corp. [2] in 1974–1975. This system utilized a DEC 11/05 minicomputer, core memory, dual cassette tape drives for program storage and data storage, and digital-to-analog and analog-to-digital (D/A and A/D) converter systems for command waveform generator and data acquisition, respectively. The system ran under an MTS enhanced version of BASIC, which allowed users to develop their own test programs. This system proved quite successful; for the first time it was possible to automatically execute fatigue tests that involved complex command histories, coupled with various on-line data acquisition protocols. The "automation" of the incremental step test (used to characterize the cyclic stress-strain curve) is an apt example of the complexity of the tests that were automated using this system. Through the use of this system it became obvious to materials test researchers that both simple and complicated tests could be performed with relative ease once the specific software (on the specific computer system) was developed; test setup and data analysis time were drastically reduced through the use of computer-controlled systems.

These positive aspects of the use of digital computers for control and monitoring of materials tests were offset by one major negative aspect—the prohibitive costs of such systems. It was simply not feasible for test laboratories with numerous test frames to allocate funds for updating each test frame with a dedicated (mini) computer system complete with central processing unit (CPU), mass storage device, terminal, machine interface circuitry, etc. Two principal methods of dealing with this problem surfaced, at least conceptually: (1) time sharing systems and (2) distributed processing systems. These two approaches are discussed in some detail later in this paper. Briefly, however, they both utilize the same idea of individual test frames sharing most major (expensive) peripherals, that is, line printers, mass storage devices, and host computer systems. The principal difference between the two systems is that with time sharing systems, the host computer is used on a time sharing basis to control and monitor the tests at each test frame. The distributed processing system requires that individual computers control each test frame independently and be linked by a communication network to the host computer system. MTS Systems Corp. opted to develop time share systems. These systems utilized DEC computers (11/34's) running under an MTS enhanced version of multiuser BASIC. The authors' personal experience with these systems proved that such systems could be useful tools for materials testing within a limited framework of application. Primarily, these systems should be classified as single-user multitask systems; the fact that each test frame controller shared time and real-time system resources with all of the other test frames being controlled dictates that each test frame controller be restricted (by the one user) on the amount of system resources that it can utilize at any given time. This meant that the test frequencies, data acquisition rates, and mass storage transfers were necessarily limited on each test frame station. The distributed processing approach, while apparently receiving considerable attention, has not (up to the present) seen any significant development by the major test systems manufacturers. The reason for this is not exactly clear, but it is the principal reason why such a system has been developed here at the University of Illinois. In a true multiuser environment, where much of the use concerns new research techniques, the time share system has proven inadequate. The distributed processing approach, which provides a real degree of independence for individual test frame users, is believed to be a solution to this inadequacy.

Recent Developments in Computing Systems

The development of electronic computing systems, at present, is probably the most dynamic field in engineering. Since the early 1950's, the speed of our largest computing systems has been increased by a factor of two, on the average, every two years [3]. At the same time, the cost (or more correctly, cost for a given level of function) of our small computer systems has fallen dramatically. This latter event is the more important from the perspective of the subject of this paper. It is the result of the development of Very Large Scale Integrated (VLSI) circuits. Integrated circuits containing close to one-half million individual components are currently in production [4]. Such circuits have the following characteristics.

1. Once in production, the unit cost of a VLSI circuit (chip) is low.

2. The nature of the technology yields devices of very high reliability.

3. VLSI circuits require relatively small amounts of power for their operation.

4. VSLI-based computers are inherently slow (about one million instructions per second) in relation to more traditionally designed computers [3].

This background sets the stage for a discussion of the organization of an automated materials testing laboratory and the computing hardware available for its implementation which follows.

Organization of Computing Hardware in an Automated Materials Testing Laboratory

The tasks which can reasonably be assumed by automated materials testing equipment are as follows:

1. Provide a smoothly functioning man-machine interface for the initiation of a test (that is, aid in specimen insertion, calibration checks, test definition, and the like).

2. Control the independent variables in a given test.

3. Monitor all test variables for out-of-range or out-of-specification conditions.

4. Capture and store all needed raw test data.

5. Detect end of test conditions and stop test.

6. Reduce raw data acquired during a test.

7. Output reduced data in man-readable form (charts, graphs, and tables).

8. Store test results in a database for future reference.

9. Provide database management tools to manipulate the stored test results.

In principle, all the requirements listed above could be met by a single computer system operating under a good real-time operating system. Such a laboratory design can be represented schematically as shown in Fig. 1. The data lines linking the central computer may carry both analog and digital data. Such a scheme has one major advantage. A relatively expensive computer can be shared by a large number of test stations. The drawbacks of the system are numerous:

1. Operation of the laboratory is dependent on one computer.

2. A heavy control and computation load on the computer from a single test frame can severely restrict the performance of all other test stations.

3. Typically, a rather large compliment of digital and analog hardware is required at each test frame.

4. If the system is designed such that the data lines carry analog information, a reduction in signal quality by transmission over long lines can result. The alternative of transmitting only digital information requires more hardware at each station.

The control scheme shown in Fig. 1. is the classic "computerized factory" model [5]. In manufacturing systems, this model has been largely replaced by some variation on the distributed system model shown in Fig. 2. As applied to materials testing, such a system partitions the nine automation system tasks listed at the start of this section into two parts. Tasks 1 through 5 or possibly 6

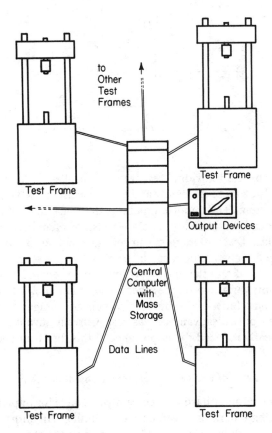

FIG. 1—*Single-processor laboratory design.*

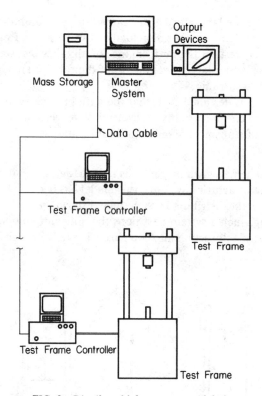

FIG. 2—*Distributed laboratory control design.*

are carried out by the individual test frame controllers. The balance of the tasks can usually be more efficiently and economically carried out by the master computer system. The important point is that the effects of failure of any computer are localized. Even failure of the master system need not stop laboratory operations. Tests in progress can continue without interruption. Such a system can be designed so data can be transmitted to a secondary computer if a test is completed and the master system is not prepared to receive data.

There are many variations on the scheme shown in Fig. 2. All of them share the common characteristic of robustness. Their advantages are compelling; historically their only disadvantage was cost. Current advances in computer design have greatly reduced the importance of this disadvantage.

Range of Hardware Choices for Laboratory Implementation

The range of computer hardware currently available for implementation of automated testing laboratories can be overwhelming. An effort will be made to classify the candidate systems and briefly discuss the strengths and weaknesses of each. No attempt will be made to provide detailed data on given systems—this sort of information is published by computer system manufacturers. While a large mainframe could be the basis for a materials testing laboratory, we will limit discussion to logical candidates.

At the top of the range in price and performance is the minicomputer. These systems are normally too costly to dedicate to a single test station. They are good candidates for the job of a laboratory master computer, particularly if the laboratory has need of a machine capable of large calculations. Good database management software can be purchased for these machines, and they can support peripherals such as high-quality plotters, line printers, and interactive graphics terminals. Incidentally, the distinction between mini- and microcomputers is becoming blurred. Many computers of this class are currently based on VLSI circuitry. In any case, if such a machine is to be purchased, be sure it can support a minimum of one megabyte of memory and that preferably it be a virtual machine (that is a computer capable of using its disk as though it were an extension of memory).

At the second level in the price performance scale, one finds machines which we shall refer to as "instrument controllers." These machines are based on the more powerful microprocessors such as the Motorola 68000. These machines can form the basis for an excellent machine controller. They are, however, relatively expensive. They normally require considerable peripheral equipment as well. They often use software well suited to the testing function (see the next section). They are less flexible from a hardware standpoint than the single-board computer systems discussed below. Such a computer can serve as a laboratory master computer, especially in a case where very large computations are not anticipated. In all other respects, they are the equals of most minicomputers. In conclusion, instrument controllers can serve as rather expensive test machine control units and make excellent laboratory master computers.

At the third level in the price scale (and essentially the same level of performance), we encounter single-board computer equipment. These machines are often manufactured by the semiconductor manufacturers and by some of the major computer manufacturers. These systems are based on the latest microprocessors. They are normally built to very high standards of quality in as much as they are intended to be components of capital equipment. These systems are supplied in a modular form so that a "semicustom" system can be configured to a given application. This modular construction allows a much higher range of flexibility than the "instrument controllers" discussed above.

Assemblies of standard single-board computer components can produce almost any level of performance desired. As a result, it is possible to achieve performance close to that obtainable by dedicating a minicomputer to a single test station. Frequently, these systems do not have their own compilers but are programmed with the aid of another computer known as a development system. Thus, software for many single-board computer systems is developed on one development system. The great power of these systems is in their flexibility. It is possible, for example, to effectively double the computing power of a single-board computer based system by adding a second single board computer to the card cage. Such an addition costs between $1000 and $2000. Multiprocessing, as this technique is called, has been known for many years [6]. The newest single-board computers make the technique straightforward as well as economically practical [7]. In conclusion, single-board computer systems make excellent test frame controllers.

At the fourth level in the price performance scale are found the so-called "personal computers." Most of these systems are not designed for instrumentation applications. Normally personal computers are built to consumer quality, not industrial quality standards. As a result they can be expected to be less reliable and less tolerant of extremes in temperature and humidity than the systems discussed above. Their software is designed for data processing applications, not real-time control. Often these systems require the use of extensive custom circuitry to adapt them to an instrumentation function [8,9]. Even so, their performance is generally below that obtainable by any of the options described above. (See, for example, the section "Comparison of Microcomputer with other Computer Control Systems" in Ref 8 where it is mentioned that the personal computer based system being described must interrupt cyclic loading of the test specimen to capture and store data.) If the system can be environmentally protected, such a computer can serve as a laboratory master system. Normally the graphical output obtainable from personal computers is not up to professional standards. The laboratory master application will be better served by the next generation of personal computers currently appearing.

Finally, it is possible to start from the VLSI component level and build specially designed systems for materials testing. The results would be similar to those achieved by the single-board computer option mentioned above. The costs of hardware development cannot usually be justified on the basis of the limited number of systems to be constructed.

In general, fast and responsive real-time computers are an asset at the test frame. Often, the premium in price paid for such hardware is more than matched by savings realized in auxiliary equipment. An example will illustrate this point. A common problem encountered in digitally driven strain-controlled tension tests of ductile materials is the large amount of relaxation observed in the plastic portion of the strain, load response. The portion of the curve in Fig. 3 just past yield illustrates this phenomenon for a strain-controlled tension test on a 10-mm-diameter bar of 1100-0 aluminum. In this test, an increment of one bit on the single board computer's 12-bit digital-to-analog converter resulted in an increment of strain of about 0.0125%. The D/A converter was incremented one bit at a time at an interval of 2.5 s. As can be seen, considerable relaxation occurs after each increment of strain.

In the part of the curve labeled "With PWM," the least significant bit was

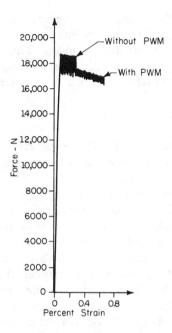

FIG. 3—*Strain-controlled tension test on a 10-mm-diameter bar of 1100-0 aluminum showing the effect of using a pulse width modulation (PWM) technique to smooth the strain command signal and reduce relaxation of the specimen. Minimum strain increment is 0.0125%. In the portion of the curve showing no PWM, the strain command was incremented 0.0125% every 2.5 s.*

repeatedly incremented and then decremented over each period of 2.5 s. On the first interval of 0.25 s the D/A converter was left at the higher setting for only 0.025 s. On the last 0.25 s interval, the converter was constantly maintained at the higher value. Between these extremes the higher output was maintained at a uniformly increasing proportion of the 0.25 s interval. The electrohydraulic test frame could not follow these short pulse changes faithfully, and as a result its response was close to a uniformly increasing strain ramp. As can be observed in the figure, the technique greatly reduced the relaxation. It is to be anticipated that a shorter pulsing interval should yield even less relaxation. Such a result can, of course, be obtained by using a much more expensive 16-bit A/D converter, but we can see that exploitation of the responsive single-board computer can yield a cost-effective solution to the problem.

Another example of the value of a relatively fast real-time computer is in peak detection. An algorithm has been written to operate on an Intel 88/40 Single Board Computer which can generate a sinusoidal or ramp command signal at a frequency up to 25 Hz and sample the response of the machine at 100 times that rate. This permits accurate detection of peak values of load and strain without auxiliary equipment. One single-board computer is able to do the job of several individual instruments.

Concentration of control and data acquisition functions in one instrument simplifies system maintenance. Basically, if something malfunctions, there is only one circuit board to exchange to cure the problem.

Software Choices

This section will concern itself with the software used by individual test machine controllers. Normally standard data processing software is adequate for the laboratory master computer.

Two schools of thought exist as to the proper type of operating environment to be used at an individual testing station. One school of thought intends to give the user maximum flexibility at the expense of considerable effort on his or her part. This scheme provides a real-time operating environment through the use of an interpreted language such as BASIC or FORTH or through the use of a real-time operating system and compiled languages such as FORTRAN, Pascal, or C. The user is required to develop his or her own testing software. Of course, over time a laboratory develops a set of standard programs used day in and day out. Vestiges of the underlying computer system remain. The user will always be running materials testing software on a real-time computing system instead of operating a computer-based materials test system.

The second school of thought places primary emphasis on the efficiency, accuracy, and ease of use of the system. The usual method of achieving these goals is through the use of menu-driven operating software often stored in read-only memory [9]. The software (often called "firmware" if stored in read-only memory) is designed to carry out a certain "class" of tests such as strain-controlled low-cycle fatigue tests. Operator setup instructions can be built into the software. Tests to ensure compliance with applicable standards are easily included. Considerable flexibility within the general test class can be accommodated. The resulting system has the feel of a very flexible and easy-to-use testing instrument instead of a computer system.

A Materials Testing System

This section describes a test frame control system currently in use at the University of Illinois. This system was conceived as a device to be used to automate a standard analog electrohydraulic test frame. The same basic hardware is currently being applied to the control of screw test machines as well. The laboratory at the University of Illinois originally was controlled by a central computer system. This arrangement proved troublesome for the reasons listed in the earlier section entitled "Organization of Computing Hardware."

The system is mounted in a standard nineteen-inch (48 cm) rack-mountable chassis 21.5 cm in height and 37.5 cm in depth which contains power supplies, memory backup battery, interfacing connectors, and a card cage. The card cage contains an Intel 88/40 Single Board Computer. This machine is configured with up to 64K bytes of read-only memory for program storage, 8K bytes of onboard read/write memory (RAM), seven 16-bit counter/timers, up to 16 channels of analog input, up to 8 channels of analog output, up to 16 digital signal lines, and two serial (RS-232C)[2] interfaces. The cage also carries an Intel RAM memory board used for data storage. This board is battery powered to protect data from power failures and can have a capacity of up to 512K bytes. (This is enough memory to store 200 cycles of load/strain data from a strain-controlled fatigue test with 100 data sets per cycle and each data value having a precision of one part in 4000.) The third board in the cage provides signal conditioning for the analog channels, battery charging circuitry, and power failure shutdown logic.

The system can be operated using a simple RS-232C compatible computer terminal. Test data are stored in the battery-protected memory and later transmitted to the laboratory master computer for data reduction and permanent storage. Data memory can be emptied and a test continued at any time; thus there is no limit to the amount of data which can be collected on a given test. The system can also be operated from a small computer such as an HP-85 if a completely self-contained control, data reduction, and data archiving system is desired.

The system was designed with two goals in mind, ease of operation and reliability.

The software is menu driven, i.e. the user is prompted to select from a menu of choices at various stages of test definition and test execution. A typical menu for fatigue testing is shown in Fig. 4a. This fatigue test menu provides for any of the following primary test functions: definition, modification, initiation, resumption, termination, and stored data display. Figure 4b shows a "submenu" of the fatigue test menu. This menu is presented when the user selects Option No. 3 of the fatigue test menu—modify test conditions. Through the use of such menus, the user can easily and quickly proceed with his particular test requirements.

On user-selectable test parameters, for example, control limits, test frequency, data sampling rates, etc., the test system issues acceptable limits of user response for the given test parameters. For example, a typical message to define (or modify) test frequency would be issued as follows:

INPUT TEST FREQUENCY
(0.00002 to 25.0 Hz)?

---

[2]Interface between data terminal equipment and data communication equipment employing serial binary data interchange, EIA RS-232C.

```
NO TEST IN MEMORY

--------- MAIN MENU ---------
INPUT AN OPTION # AND <CR>
1=) FATIGUE TEST OPTIONS
2=) CALIBRATE A/D, D/A
3=) TRANSFER DATA TO HOST
4=) CLEAR DATA STORAGE AREA
5=) PERFORM MEMORY TESTS
6=) MODIFY HOST COMMUNICATIONS
?1


---- FATIGUE TEST MENU ----
INPUT AN OPTION # AND <CR>
1=) DEFINE A NEW CYCLIC TEST
2=) START/RESTART CYCLIC TEST
3=) MODIFY CYCLIC TEST
4=) DEFINE RAMP PROFILE
5=) RUN RAMP PROFILE
6=) DISPLAY STORED DATA
7=) RETURN TO MAIN MENU
8=) END TEST
--- TYPE "S" TO HALT TEST --
?
```

FIG. 4a—An example of a main option selection menu.

```
**** TEST MODIFY MENU ****
INPUT AN OPTION # AND <CR>
1=) FEEDBACK MODE
2=) WAVEFORM
3=) CONTROL LIMITS
4=) CONVS PER CYCLE
5=) DATA STORING SCHEME
6=) FREQUENCY
7=) UNDERPEAK DETECTOR
8=) FAILSAFES
9=) STOP COUNT
10=) RETURN TO TEST MENU
?4

INPUT DESIRED NUMBER OF DATA
SAMPLES PER CYCLE (50 TO 500)
?200


INPUT TEST FREQ. IN HZ.
  0.000004 TO  6.25000000E+000
?10
  0.000004 TO  6.25000000E+000
?6
```

FIG. 4b—An example of a submenu (obtained in this case by choosing Option 3 in Fig. 4a).

Any operator input outside of these limits would not be accepted; the user would be prompted to reinput the test frequency until the frequency was within the specified limits. Not only are the limits on individual test parameters checked as they are input, but also limits on interdependent test parameters are set and checked during test definition (or modification). For example, the actual upper limit of attainable testing frequency is dependent on the user-selected data sampling rate and is calculated using this parameter. This is one clear-cut advantage to a menu-driven software system—the overall interaction of (worst case) test conditions can be checked out (and controlled) in the computer system development laboratory instead of the mechanical test laboratory. The software is designed to allow the user to easily interact with the test machine. During a test run, the test can be stopped at any time. In the event of a power failure, care is taken that the test specimen not be damaged. After power is restored, all data will be intact due to the battery backup provision. The test can be restarted from the point at which it was stopped in a matter of seconds.

Besides the basic requirements of the software being easy to use and highly reliable, another area of software operation that has been given considerable attention is that of thorough test documentation. When a fatigue test is initially defined, all pertinent test parameters are stored in the battery backed RAM, memory. Furthermore, as modifications (if any) are made to the test parameters, these modifications are also logged into the data storage area. Even the ramp segments (which may optionally be defined and executed any time the cyclic test has been temporarily suspended) are always logged into the data storage area. Thus, the data storage area is used to completely store the history of the test as actually performed as well as to store conventional (user selectable) data sets, for example, cycles of stress-strain data for a low-cycle fatigue test. This philosophy of thorough test documentation is considered relevant to standards groups such as ASTM. It is clearly a superior method of test documentation than a simple valid/invalid label tagged to the test based on any group's *current* criteria of what constitutes a valid test. With this in mind, no attempt has been made for the system to force the user to comply with any sets of standards in the definition of a test; however, anyone who may wish to use the results of the test can decide, using an appropriate criterion, whether or not the test has validity within the framework of his or her particular analytical needs.

Overall reliability is assured by three factors. One factor is the battery-powered memory described above. A second factor is the use of high-quality hardware throughout. The third factor is the very simplicity of the system. In particular, troublesome flexible disk storage units are avoided. Hardware problems are minimized and operator errors nearly eliminated due to the system design.

The system assumes the use of a second computer for data reduction and data archiving. This permits the optimization of the controller for the test control and data acquisition function. Data reduction and long-term storage are easily and efficiently carried out using standard data processing hardware and software.

The basic system has great flexibility. While the present system is designed to interface to a standard electrohydraulic test frame using standard analog servocontrol, a second single-board computer could be added to the system to take over the servocontrol function, and incidentally permit virtually any control law desired. A second area of flexibility is in software design. The software is extremely modular. It is, in the main, written in Pascal (specifically the H-P 64000 dialect of Pascal). Only a few time-critical procedures are coded in assembly language. Even these assembly language procedures have Pascal prototypes. Initial software was written to carry out low-cycle fatigue tests on an electrohydraulic test frame. Currently, this software is being extended to conduct creep fatigue tests on either an electrohydraulic test frame or a stepper motor actuated screw test frame. Software for the screw machine and the electrohydraulic machine is nearly identical—differing only in actual machine drivers. A final area of flexibility is in communications. While the present system is designed to communicate with a laboratory master computer via a serial RS-232C interface, many other communication options exist. Some of these options will be discussed in the following section.

Data Communication

The distributed nature of the testing systems discussed in this paper places heavy demands on data communications. The entire area of digital communications is currently in a state of rapid change. This section is by no means intended to be a complete review of the state of the art, but is rather intended as a guide to those communications standards and schemes most useful in the testing laboratory. One of the most widely used digital communications avenues is based on the serial RS-232C standard. This is an electrical standard which usually implies the use of the ASCII character code.[3] Most modern computer terminals use an RS-232C serial interface and operate on ASCII coded characters. As a result, the great majority of computer systems have provision for the attachment of such a terminal. A simple means of communicating with such a computer involves making the test frame controller appear to be a terminal to the master computer system. Data can simply be input to the master computer using its keyboard data entry software. That is, the master computer is "tricked" into acting as though an individual were entering the data manually from a keyboard. Data transmission rates are usually adequate for the quantity of data collected in material tests such as low-cycle fatigue. (Usually the maximum data rate is 19 200 baud, which means about 1920 characters per second.) Advantages of this communication mode are its universality and the fact that data can be transmitted over long distances using modems (virtually to anywhere in the world reachable by telephone line). One disadvantage of the system is the fact that each test frame controller requires a separate cable to the master computer and either its own port on the master or a multiplexer to switch a given controller to the master.

A somewhat newer communication standard is RS-422.[4] One particular version of this standard—the multidrop network—is especially attractive for

---

[3]Code for Information Interchange, ANSI X3.4-1977.
[4]Electrical characteristics of balanced voltage digital interface circuits, EIA RS-422A.

a testing laboratory. The laboratory master computer acts as the network master. The test frame controllers are attached to a single cable which links all test stations (slaves) to the master computer. The network is so designed that all slaves monitor data coming from the master, but only one slave is permitted to transmit at a time. It is the master controller's responsibility to select a slave to "talk." Slaves are not permitted to transmit data directly between themselves. This is not a serious limitation for a testing laboratory. This is a serial data transmission interface, basically an outgrowth of RS-232C. Maximum data rates are comparable to those achieved using RS-232C; however, having sacrificed hardware universality, a sacrifice in software universality can greatly increase data transfer rates. The usual method of transmitting serial data is via ASCII character code. Thus the number -3572 is transmitted as five separate characters. The information in the number -3572 can be transmitted in the equivalent of only two characters. As a result, the use of special software running on the master station can reduce data transmission time by more than 50%. Other benefits of the multidrop network include the ease in which it is possible to monitor or modify factors such as test station status and test parameters from the master computer.

Another communication avenue of some interest is the instrumentation bus IEEE 488-1978,[5] sometimes known as GP-IB. This is a bus system which transmits data in parallel, eight bits at a time (byte serial). Up to 15 devices can be interconnected on the bus. The data rate is quite high, normally in the region of 50 000 bytes per second (an effective baud rate of 500 000 bits/s). Control of the bus can be passed from one device to another. Normally any device on the bus can be made to communicate with any other device on the bus. The only real limitation on the system is distance. Normally two devices on the bus should not be spaced more than 3 m apart. The total bus length should not exceed 20 m. High-speed bus extenders have become available which can operate up to a distance of 1000 m, but they are rather expensive. A low-speed bus extender is also available which operates at serial data rates, but it is both slow and expensive. It is felt that IEEE 488 should be considered a candidate for instrumentation linkage at the individual test station. Many standard instruments such as voltmeters, signal generators, and data acquisition systems are now available with IEEE 488 interfaces. Most "instrument controllers" discussed above come equipped with this interface. One advantage of using single-board computer hardware at a test station is that an IEEE 488 interface can be added by purchasing and installing a simple plug-in module.

One final communication medium will be discussed. This medium is the "Xerox Ethernet-like" version of the IEEE 802[6] networking standard cur-

[5]Institute of Electrical and Electronic Engineers (IEEE) Standard Digital Interface for Programmable Instrumentation, IEEE Standard 488-1978.
[6]CSMA/CD access method and physical layer specifications, provisional IEEE Standard P 802.3 Draft D.

rently reaching the final stages of approval. This scheme is primarily intended for communication between large computer systems. A complete 802 interface is projected to cost about $800 even when it is mass produced. The interface also requires considerable software overhead. As a result, one would probably consider 802 as a means of linking the laboratory master computer to a central computer facility. One interesting factor, however, is that the fundamental VLSI circuits needed to implement the 802 standard will probably become quite inexpensive within the next year or two. As a result, a nonstandard custom network based on these chips and using only the features of 802 needed for the laboratory communication function could be used to build a very efficient laboratory data communication system. One advantage of this communications mode is the fact that data are transmitted on a simple coaxial cable. A second advantage is the very high data rates obtainable, comparable to those obtained using IEEE 488.

Impact of Current Advances in Testing Methods on Testing Standards

The dedication of a computer to the individual test station opens new opportunities for the control and measurement of materials tests. One of the tasks which the computer can carry out is assuring that independent parameters truly match their specified values. For example, the test station controller developed at the University of Illinois is capable of measuring the applied strain peaks and strain mean during a strain-controlled fatigue test and adjusting the strain command signal until specific peaks and means are matched to within a user-specified tolerance. This feature is particularly useful for higher-frequency tests where the dynamic characteristics of the servohydraulic test frame become important. Another task which a test controller handles well is documentation of test history. The systems advocated in this paper have the capacity of virtually unlimited data storage. While this feature should not be abused, it does permit tracking of the entire test history of a given specimen.

Materials testing standards for automated tests should be written such that both the requirements for test quantification and the capabilities of automated testing methods are considered. In particular, the impulse to measure a large number of parameters "because we can" should be avoided. Such measurements are not free. In the same vain, extremely tight specifications on measurement tolerances yield no real gain in information, yet require much more costly instrumentation.

An example of this can be illustrated as follows. Suppose the required accuracy on load cell calibration for a particular test standard is set at $\pm 1.0\%$ of full scale. The overall (worst case) error on the calibration on an automated test system would be the sum of (worst case) error of the (analog) calibration and the (worst case) error of the analog-to-digital converters on the automated system. For a 12-bit-resolution A/D converter, this error should be on the order of ±0.05% (that is, 1 part in 2000 assuming perfect calibration of the A/D converter). Clearly, this error is almost negligible compared with ±1.00%, and hence any attempt to provide improved A/D resolution (through say a 14- or 16-bit A/D converter) would be ill-advised for the particular test being considered. Basically, test standards should be written such that all independent variables are monitored and preferably controlled within reasonable tolerances. Enough data should be recorded to permit reconstruction of the test. In case of doubt, the error should be on the side of taking excess data, but every effort should be made to avoid needless data storage.

One final recommendation comes as a result of the well-known aliasing phenomenon [10]. Digital sampling systems are prone to this problem. A simple illustration involves sampling a 100 Hz sine wave signal of 5 V amplitude at a sampling rate of precisely 100 samples/s. Depending on when the sampling starts relative to the sine function, the signal will appear to be a constant signal with a magnitude somewhere between −5 and +5 V. The effects of aliasing are avoided if sampling is always carried out at, at least twice the frequency of the highest frequency component of the signal being measured. Analog guard filters are normally specified to eliminate high-frequency interference signals from analog data signals before they are digitized. One difficulty in specifying such filters for materials test applications is the extremely wide range of test rates commonly encountered. If the filters are not to attenuate test data from high-rate tests, they must be set to such a high break frequency that common interference sources such as 60 Hz power line interference can potentially cause aliasing of data taken at a slow sample rate. One costly solution to this problem is tunable guard filters. A second solution is to insure the absence of interference which might cause aliasing. This can be accomplished through careful analog instrumentation practice. Some test of the success of such practice is advisable. One technique would involve sampling a constant input signal at successively higher sampling rates until reaching a sampling frequency about four times the break frequency of the input guard filter. Clearly no appreciable change in the measured data should be detected.

Conclusion

This paper has endeavored to describe current trends in automated materials testing systems with particular reference to the system being developed at the University of Illinois. The recent startling developments in VLSI circuit technology will lead to substantially improved materials testing systems in the very near future.

References

[1] Richards, F. D. and Wetzel, R. M., *Materials Research and Standards Magazine*, Feb. 1971, pp. 19–22 and 51–52.

[2] Donaldson, K. H., Jr., Dittmer, D. F., and Morrow, J. in *Use of Computers in the Fatigue Laboratory, ASTM STP 613*, American Society for Testing and Materials, Philadelphia, 1976.

[3] Levine, R. D., *Scientific American*, Vol. 246, No. 1, Jan. 1982, pp. 118-135.

[4] Patterson, D. A., *Scientific American*, Vol. 248, No. 3, March 1983, pp. 50-57.

[5] Kahne, S., Leflcowitz, I., and Rose, C., *Scientific American*, Vol. 240, No. 6, June 1979, pp. 78-90.

[6] Dijkstra, E. W., *Communications of the Association for Computing Machinery*, Vol. 8, No. 9, Sept. 1965, p. 569.

[7] Adams, G. and Rolander, T., *Computer Design*, Vol. 17, No. 3, March 1978, pp. 81-89.

[8] Fleck, N. A. and Hooley, T., "Development of Low Cost Computer Control," *Proceedings*, SEECO '83, International Conference on Digital Techniques in Fatigue, The City University, London, B. J. Dobell, Ed., 28-30 March 1983, pp. 309-316.

[9] Barker, D. and Smith, P., "A Micro-Processor Controller for a Servo-Hydraulic Fatigue Machine, *Proceedings*, SEECO '83, International Conference on Digital Techniques in Fatigue, The City University, London, B. J. Dobell, Ed., 28-30 March 1983, pp. 279-290.

[10] Blackman, R. B. and Tukey, J. W., *The Measurement of Power Spectra*, Dover, New York, 1959, p. 31.

*Dale A. Meyn,[1] P. G. Moore,[1] R. A. Bayles,[1] and
P. E. Denney[1,2]*

An Inexpensive, Multiple-Experiment Monitoring, Recording, and Control System

REFERENCE: Meyn, D. A., Moore, P. G., Bayles, R. A., and Denney, P. E., "**An Inexpensive, Multiple-Experiment Monitoring, Recording, and Control System,**" *Automated Test Methods for Fracture and Fatigue Crack Growth, ASTM STP 877*, W. H. Cullen, R. W. Landgraf, L. R. Kaisand, and J. H. Underwood, Eds., American Society for Testing and Materials, Philadelphia, 1985, pp. 27–43.

ABSTRACT: Strip chart recorders and data loggers have numerous shortcomings for monitoring sustained-load cracking (SLC) and fatigue tests, principally because they record at fixed rates. During a long test this affords low resolution during rapidly changing events and creates large amounts of mostly useless data. For over two years the authors have used a personal computer to monitor tests and record data for SLC and fatigue experiments. The computer stores specimen identification and test parameters, converts raw data into usable form, displays current test status, notes and acts on various test status parameters, periodically stores significant data on a floppy disk system, and automatically terminates tests as specified by the operator. The decision-making capability of the computer greatly reduces the amount of nonsignificant data to be stored while permitting more rapid data acquisition when the signals begin to change rapidly. A fast-switching battery-inverter system provides standby power for up to two and a half days in the event of power failure. If both primary and standby power fail, the computer's autostart feature allows it to resume data collection when power resumes.

The use of BASIC permits software to be produced in-house and allows revision as operational needs change. The input routines, start-up, and running of experiments are completely interactive, and designed to prevent omissions and errors by the operator. Presently four experiments can be conducted simultaneously, but a newly acquired 16-channel, 12-bit analog-to-digital converter will allow for considerable expansion. Because system response is inadequate to measure and record fatigue load signals directly, they are filtered to produce an average value which is recorded. The analysis of data stored on disk files is done using another personal computer to avoid interference with data acquisition. A unique feature of the analytical method is the use of the decrease in load with increasing crack length in the stiff, displacement-controlled test configuration to calculate crack lengths, instead of using a clip gage across the notch.

KEY WORDS: computers, microcomputers, computer interfacing, BASIC programming, automation, mechanical testing, data acquisition, sustained load cracking, fatigue

[1]Naval Research Laboratory, Washington, DC 20375.
[2]Presently, Westinghouse Electric Corp., Pittsburgh, PA 15222.

The acquisition and recording of load and other experimental parameters as a function of time are basic to a wide range of mechanical experiments, and many techniques are currently in use. The most common methods employ strip chart recorders or data loggers, both of which record data at a fixed rate. Usually, the recorder chart drive is set to a compromise speed which will preclude running out of paper before the end of the experiment while providing some resolution of rapid events, such as the rather rapid crack growth occurring at the end of a sustained-load cracking (SLC) experiment. Data loggers are similarly set to provide a compromise between available recording space and data resolution. The mechanical problems associated with the servomechanisms, pens, and paper transport in strip chart recorders are notorious, and both recorders and data loggers require transcription of data for subsequent processing, a process fraught with misery and error.

Microprocessor- or microcomputer-based systems have been used to collect data and to control experiments [1,2], and can offer advantages over conventional approaches. A microcomputer can be programmed to select only significant data, thus reducing the flood of information to the manageable essential. The data can be stored directly in computer-accessible format, simplifying subsequent processing and presentation, and reducing the chances for introduction of error caused by manual transcription and processing of data. Furthermore, a microcomputer program can control and monitor the experiment, reducing the need for human intervention and allowing experiments to be safely and effectively conducted at night and on weekends.

The availability of cheap general-purpose microcomputers with virtually minicomputer capabilities, and abundant plug-in converters and peripherals for them at very low prices, has in the past five years or so completely changed the approach to home-brew automation of experiment control and data collection. Fast multichannel analog-to-digital converters of high resolution with built-in driver programs, magnetic disk memory systems, and fast dot matrix printers are now available for the hobby and small research device market. These new, cheaper devices coupled with much friendlier programming languages and computer operating systems enable anyone with a hobbyist's interest and tenacity to set up a workable system and write his own programs. The result might well horrify systems designers and program analysts, but can be satisfactory nonetheless. The primary advantages of the home-brew approach are that the system is completely under the experimentalist's control, the user can modify software and hardware personally at any time, and the direct financial outlay is very low. Several thousand dollars should suffice for everything, including duplicates of all critical components, even the microcomputer. Finally, it is not easy to find a commercial system tailored to the specific needs of certain kinds of tests, especially at low cost. Quite often, such systems are complex, expensive, and rather general purpose, making dedicated use difficult to justify.

This paper describes a system which uses a popular consumer-type micro-computer with plug-in peripheral and interfacing devices to control, monitor, and record data from sustained-load crack propagation and fatigue experiments. The emphasis in the development of this system was on reliability, accuracy, and versatility at minimum cost.

System Description

Hardware

The heart of the system is an Apple II+ personal microcomputer, which has several multipin accessory slots into which are plugged the auxiliary and input-output devices that allow the computer to accept and store data, keep track of time, and monitor and control the various experiments. The computer has a total of 65 536 bytes (1 byte = 8 binary digits and represents 1 character) of addressable memory, of which approximately 36 500 bytes are read-write memory available for user programs and temporary data storage. The latter is volatile memory; that is, its contents disappear when the computer is turned off, so two 5¼-in. (133 mm) mini-floppy disk drives are attached to the computer through a disk drive controller card plugged into one of the multipin accessory slots to provide for permanent storage of the disk operating system (DOS), user programs, and data on thin (floppy) magnetic recording disks. When the computer system is switched on, the disk system is automatically "booted" (its operating system program is loaded into memory), and the experiment control program is loaded into memory and run. The primary display for the computer is an industrial TV monitor, and manual input is accomplished via a 66-key typewriter-style keyboard. An internal speaker can be controlled by the program to alert the operator to improper keyboard entries, system malfunctions, and to the occurrence of significant test events, such as specimen fracture.

A quartz-crystal-controlled electronic clock system (Mountain Computer, Inc.) is plugged into an accessory slot to provide time information for load-versus-time recording and for controlling the sequencing of data acquisition and storage. The clock has millisecond resolution, is accurate to 0.001%, and has 388 days' duration before its counters must be reset. The clock readout is under control of the operating program, but the clock oscillator and counters are powered independently of the computer and continue functioning if the computer is turned off. A rechargeable standby battery on the clock card provides up to 4½ days of operation even if all power to the clock is interrupted.

Digital computers cannot directly accept ordinary electrical voltage-current information such as is provided by a load cell signal conditioner/amplifier (LCCA) or other conventional transducer systems. Such analog signals must be converted to binary digital codes by an analog-to-digital converter

(ADC). An Interactive Structures, Inc., Model AI 13 12-bit (1 bit = 1 binary digit) ADC with an integral 16-channel input multiplexer (channel-selecting solid-state electronic switch) is plugged into one of the computer accessory slots. The conversion from analog voltages to digital numbers which the computer can accept requires only 20 μs per channel. The 12-bit capability implies digital resolution of 1 part in 2^{12}, or 0.025%, which is ample for mechanical testing. This ADC replaces a more complex system consisting of a slow-conversion, single-channel 12-bit ADC fed by a laboratory-made 4-channel solid-state multiplexer which was in turn controlled by analog signals from the DAC (digital-to-analog converter) section of a high-speed Mountain Computer, Inc. 8-bit 16-channel DAC+ADC plugged into a separate accessory slot under control of the operating program. In practice, the newer 12-bit 16-channel ADC and the old hybrid system operated in much the same way from the computer programming point of view, but the old system was much slower, requiring over 500 ms per conversion, and enabled simultaneous operation of only four experiments.

The 8-bit 16-channel DAC+ADC also provides two-way communication between the computer and the various experiments. It allows the operating program to turn motors, signal lights, and other devices on and off through relays, and permits the computer to sense the status of various experimental devices, to sense power failure, or, if one desires, to monitor and record data at low resolution (approximately 1%).

The loads were converted to electrical analog signals in the conventional way, using strain-gage load cells and Measurements Group Model 2100 load cell conditioner/amplifiers. It was necessary that the LCCAs used be electrically compatible with the input circuits of the ADC system. This is made relatively simple by the use of LCCAs having continuously adjustable gain and high-stability, low-residual ripple amplifiers. Early experiments showed evidence of large electrical system voltage transients entering the signal circuits and providing false data. Their effect was reduced by simple RC (resistance-capacitance) filters in the signal circuits. Additional filtering was needed for fatigue tests, to convert the sinusoidal (ac) voltage-time waveform produced by the sinusoidal load to a static (dc) signal representing the mean load. Fatigue cycles are counted using an Interactive Structures, Inc. Model DI09 counter/timer interface.

The entire computer system (except the TV monitor) plus the LCCAs are fed from a standby power system (Welco Industries Model SPS-1-250-12) which incorporates a fast a-c power sensing circuit, a relay, 110 V a-c/12 V d-c inverter/charger, and a large 12 V d-c lead-acid battery. Normally the computer system is fed directly from 110 V a-c mains, but if the sensor detects power failure, the system is switched to ac from the inverter powered by the 12 V d-c battery, for up to 2½ days if necessary. The switchover is so fast that no interruption is sensed by the computer.

Software and Operating System

The software used to monitor and control the experiments was developed in Applesoft, an enhanced version of the BASIC language. BASIC is an interpreted language, and when changes are made in the progam the result can be immediately run without compilation. The language is simple and similar to English, yet very flexible and powerful in its enhanced version. These advantages outweighed the greater speed of compiled languages such as FORTRAN for this application. The software discussed here has evolved as deficiencies were recognized and corrected, and contains many features intended to prevent or alleviate the effects of bad operator habits, to sense predictable component failures and deal with them, and to make operation as simple and "fail-safe" as seems reasonable. The operator input routines make use of default parameters which appear on the input line on the display screen to be accepted or changed as desired. This feature is especially useful when one makes one or a few corrections to a long list of entered items and reduces input errors considerably. Given the great flexibility of the microcomputer operating system and expanded BASIC programming features, the process of refining the software could become endless, a possibility which had to be assiduously guarded against.

The software is composed of several program blocks. A flow chart in Fig. 1 shows the blocks and their relationships. This program is automatically loaded from disk into computer memory and run whenever the computer is powered. It first initializes the system by setting up numerous parameters and tables used by the program, then searches the disk storage system for a list of experiments in progress. If none are in progress, control goes to the menu, which waits for operator selection of an item on the menu list. If one or more are in progress, the appropriate disk data files are examined and necessary information (specimen identification, specimen parameters, experiment status, etc.) is put into computer memory, and control is transferred to the experiment scanning and data collection (DATASCAN) routine, Fig. 2.

The menu is normally the starting point unless the program has restarted after a power outage or because the operator wished to temporarily suspend operation for some reason. The menu options are indicated in the flow chart by arrows pointing from the menu toward the blocks. The first option used will normally be "Start an Experiment," and related to this are "End an Experiment" and "Change an Experiment." These options, if all goes as the operator desires, automatically transfer control to the DATASCAN loop, but the operator has the option of aborting to the menu if things do not seem right. From the menu one can also choose the diagnostic options (which return to the menu on completion) or go directly to the DATASCAN loop (given any active experiments), or exit the program.

The option "Start an Experiment" leads the operator through a series of

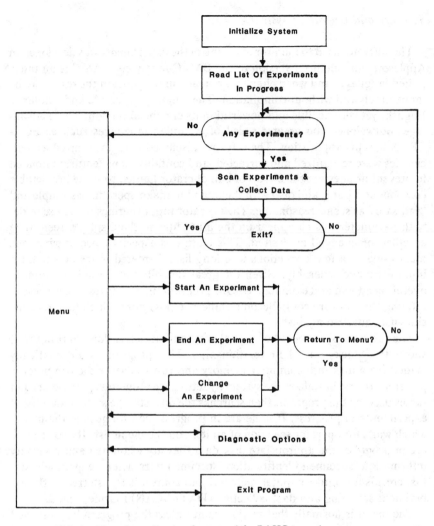

FIG. 1—*Flow chart of major elements of the BASIC operating program.*

instructions in question-and-answer format (using the default input parameter concept) which ensures that all necessary information about the type of experiment (SLC or fatigue), specimen type and dimensions, and experimental parameters (desired stresses, etc.) is entered into memory via the keyboard. The routine calculates required load parameters, checks to see that load cell ranges are not exceeded, and directs the operator to properly adjust the LCCA load ranges. After all information is stored in data files on disk, the routine leads the operator through the steps necessary to properly load up the specimen and start the experiment. If all goes well, control automatically

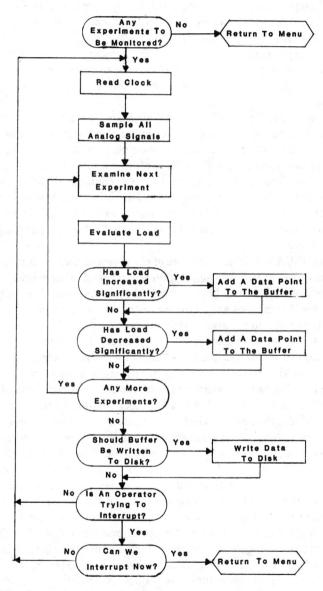

FIG. 2—*Flow chart of the experiment scanning and data collection (DATASCAN) subroutine.*

transfers to the DATASCAN routine when load-up and machine adjustments are completed. Otherwise, the operator can at any time elect to abort and return to the menu.

The "Change an Experiment" option allows reentry to a point in the "Start an Experiment" routine where loads and machine adjustments can be changed, usually to increase the load if nothing is happening, especially during an SLC experiment. The "Stop an Experiment" option allows manual termination of an experiment by removing its number from memory and from the "Active Experiments" disk file, and it sends all data still in memory to disk storage.

The "Diagnostic Options" lump together routines intended to allow the operator to look at the data on any given channel in some detail, and to allow recovery from disk failure. The latter routine permits the operator to copy the necessary information from a good disk onto a new disk to replace a bad disk, or if both disks have failed, to reconstruct the contents of the storage disks from information in computer memory. This information will include the program and experimental and specimen data, as well as load-time data. The latter data will be incomplete, consisting of initial values plus any values currently in computer memory.

The DATASCAN routine is the core of the program because it monitors the experiments and controls the acquisition of data, and other routines transfer control back to it after they have performed their function. The routine first checks to see if there are any active experiments, as denoted by a nonzero specimen number(s) in the appropriate memory location. It also checks to see if the power is on to the fatigue machine motors and LCCA power supplies for the active experiments. If anything is "alive," it reads the clock, then directs the 16-channel ADC to sample and register the signal voltages from the active experiments. This latter process requires about 20 μs per channel, and is practically instantaneous, so that a single clock reading suffices for all channels. The old single-channel ADC plus separate multiplexer worked somewhat differently, requiring a separate clock reading for each channel because of the slowness of the ADC acquisition, which required about one *second* per channel. Next, a loop is entered in which the digitized voltage reading for each channel in turn is compared with the value last stored as a data point for that channel. If a significant change has occurred in any channel, a new data pair representing the load and time or number of cycles is stored. The dwell time per channel for comparison and decision-making is about 250 ms, and if a significant load change has occurred, a further 5 ms delay is associated with storing the data point. For display or subsequent processing, the raw load data are converted to the actual load values using calibration coefficients which are also stored for each experiment. After all the active channels have been read, the DATASCAN routine checks to see if an interrupt has been requested from the keyboard or by the computer memory-to-disk data transfer routine. If so, and no load changes have occurred recently (that is, DATA-

SCAN is not "busy") the interrupt is handled. Otherwise the DATASCAN sequence begins again, unless the operator elects to override this safety feature.

The selection of data by the DATASCAN routine is based on the need to ensure acquisition of significant data, and reduce the acquisition of incidental noise and transients not related to the experimental results. This is done primarily by careful prior selection of the signal voltage change, ΔV, which will be accepted as representing a significant change in load from the previously stored value. Under the selected experimental conditions, the recorded parameter—load—decreases monotonically with time as the crack grows, or stays level if no crack growth occurs. Any increase in load is a result of temperature variations, electrical transients, amplifier drift, etc. Such effects can cause a very large number of false data points to be recorded, and the smaller the value of ΔV, the greater this number. It is necessary to select a value of ΔV large enough to reject small fluctuations, but small enough to provide sufficient sensitivity to realistically record the data trend. A digital filter is used, whereby three unevenly spaced readings taken 6.5 and 8.5 ms apart are compared. A mutual disparity exceeding 0.2% results in rejection of the readings. This is a very effective noise rejection procedure. In order to further decrease the sensitivity to false signal changes, the value of ΔV for acceptance of positive signal excursions is made larger than for negative excursions which correlate with decreasing load. This tends to ratchet the data in the negative direction, but in practice it has proven satisfactory in preventing overloading of data storage capacity by electrical transients and occasional periods of instability in amplifiers and other devices in the signal path. It would of course be preferable to have absolutely noise-free, stable measuring and amplifying circuits, but this ideal is very difficult (possibly not feasible) to obtain. Experiments can be terminated automatically upon reaching some operator-designated end-point, or manually as the operator desires. In SLC tests the experiment is scanned until the operator notices that the desired end-point, such as fracture, has been reached, whereupon he requests a exit from the DATASCAN loop to end the experiment. Automatic shutoff is not needed, because, as will be described, no active machinery is used to maintain load. Fatigue experiments, on the other hand, benefit from automatic shutoff to save the motors from unnecessary wear and the fractured specimens from flailing about and ruining the fracture surfaces. Shutoff occurs when the mean load drops to a percentage of initial value determined by the operator. Automatic shutoff presents the risk that a transient loss of signal will terminate the experiment, but this can be recovered by the operator without significant harm since the fatigue motors will merely stop and leave the specimen at some static load. This automatic shutoff feature can also be used to fatigue-precrack specimens because the shutoff load is related to crack depth, and this relation can be calibrated with reasonable accuracy for most purposes.

The transfer of data from volatile computer memory to permanent magnetic disk storage via the disk drive and its disk operating system (DOS) is a most important feature. Two disk drives are used to provide duplicate storage. If one of the drives fails to operate properly or the disk is damaged by dust particles, etc., the DOS can usually detect trouble via an error-checking system. The program continually looks for such errors and, if one is detected, alerts the operator via visual and audible signals and avoids the bad disk or drive. If both drives fail, the alarm signals become more intense. In either case, options are available in the program via its menu to repair the effects of partial or total disk failure to the greatest possible extent. The transfer from computer memory to disk is activated when the program judges that enough time will be available to complete the transfer before another significant load change will occur, *and* enough data are in memory, or it has been a long time since such a transfer has taken place. Typically, if it has been several minutes since the last significant load change occurred (all channels considered), there are several dozen data pairs, or it has been 24 h since the last transfer and at least one data pair is in memory, the transfer takes place. Transfer from memory to disk also occurs whenever the operator interrupts the DATA-SCAN routine regardless of other status factors.

An important safety factor in the DATASCAN routine is invoked if power is interrupted. Normally, the standby battery-powered inverter system takes over. However, the fatigue machines will not continue to run, as they use too much power for the battery to sustain very long, and are connected only to main power. The program senses motor stoppage and suspends the fatigue experiments while continuing to collect data from the sustained-load experiments. If for some reason the LCCA power supplies cease functioning, all experiments are suspended and the operator is alerted. Upon return of power, in either case, normal operation resumes. If the computer system is shut down, the fatigue machines will be stopped, and of course no data can be recorded. When power resumes, the information stored on disks is fed back into the computer after the autostart operation described earlier. The program then resumes collecting data as if no interruption has occurred, although there will be a gap in the data if anything has happened to an SLC experiment (which maintains load during power failure). The fatigue experiments are restarted and usually will not have suffered significantly from the hiatus, since the motors will have stopped upon power failure in any case.

Processing of the raw data is done off-line on a separate, identical computer system (without ADCs, etc.) to minimize interruption of the data gathering system. In practice, the operator signals the DATASCAN routine to avoid attempting to perform disk storage, removes one of the data disks from the drive, copies the desired data files onto another disk using the other computer system, returns the data disk, and resumes normal operation. This takes a few minutes, so it should be done during periods when the experiments are not very active even though data will still be acquired in memory. Whenever an experiment is completed, all its data are copied from the data disk to archive disks, then erased from the data disks to free space for other experiments.

Examples of Results

Experimental Methods

SLC Experiments—Sustained load cracking (SLC) experiments were conducted using modified compact tension specimens which were approximately 52 mm square by 26 mm thick (2.08 by 1.04 in.) with 2.6 mm (0.104 in.) 60-deg V-shaped side grooves [3]. They were precracked in fatigue to $0.3 \leqslant a/W \leqslant 0.5$ ($a = 5$ crack depth, $W = $ width, measured from loading holes' centerline). The materials were α-β processed and β-processed Ti-6Al-4V, annealed at 1200 K for 7 h, then furnace-cooled. The specimens were loaded in a very simple rectangular steel frame of high stiffness, by threaded 25-mm (1 in.) steel rods passing through the ends of the frame [4]. Load was measured by a load cell in series with the specimen, and applied by manually wrenching a large nut on the protruding rod end. This configuration causes the load on the specimen to decrease as the crack grows, because of resulting increase in specimen compliance (compliance is deflection per unit load) [4]. It is a simple matter to calibrate this crack length versus relative load drop relationship and use it to calculate crack depths from load versus time data [4]. In principle this is identical to the use of clip gages [5] to obtain crack depth information, differing mostly in being less sensitive to small changes in crack depth.

Fatigue Experiments—Fatigue life data were obtained on β-processed Ti-6Al-4V sheet materials using small, upright-cantilever bend specimens [6]. The specimens were of various sizes and geometries to accommodate the material available and the needs of the particular experiment. Results to be presented were obtained on 12 mm (0.48 in.) square bars 120 mm (4.80 in.) long, having cylindrical cross section waists machined near one end [6] to concentrate the bending stress at a well-defined location. The radius of curvature at the waist was 12 mm (0.48 in.), and the minimum diameter was 6 mm (0.24 in.). Some of the specimens were tested as received; others had been laser surface-melted to see what the fatigue properties of rapidly solidified Ti-6Al-4V would be. Loading was accomplished by pulling the free end of the specimen using a small electric motor, a pulley drive coupling, and an adjustable crank. The stiffness of the drive system results in displacement-controlled fatigue, such that as the crack grows, the load decreases. This allows the decrease in load sensed by a load cell to be used to evaluate the progress of the experiment. A load cell in series with the pull rod was used to sense load. No systematic calibration of load drop versus crack length was developed, but such calibration should be possible. This system was also used to precrack small cantilever beam stress-corrosion crack test specimens by noting the load drop and stopping at a preselected load.

Experimental Results

Sustained Load Cracking—Figure 3 shows typical load-versus-time data for an SLC test conducted on a β-processed specimen in an alcohol bath at 196 K. The resulting crack growth rates for this specimen (No. 667) are shown in Fig. 4. The load values were converted to crack depths using the load drop calibration procedure, and the unsmoothed crack depth versus time data were converted directly to crack growth rate data by dividing increments of crack growth by corresponding time intervals. The crack depth increments used were 0.75 to 1.5 mm (0.03 to 0.06 in.). The initially high crack growth rate is due either to initial transient creep in the cracked specimen, or to initial reshaping of the originally nearly straight fatigue crack into the more bowed-out configuration of the stable SLC crack front.

The crack depth-load drop calibration approach worked satisfactorily, giving reproducible results, although not with the sensitivity to small increments of crack growth possible with crack mouth displacement gages. SLC measurements conducted with two identical specimens (Nos. 383 and 384), cut from a single piece of α-β processed Ti-6Al-4V vacuum dehydrogenated at 1200 K and furnace-cooled, showed very good agreement in crack growth rates between specimens, Fig. 4. The low hydrogen contents achieved (8 ppm weight) and uniformity in microstructure between specimens reduced the scatter from metallurgical factors, which in most SLC tests was discouragingly high. This set of measurements was made using strip chart recorders,

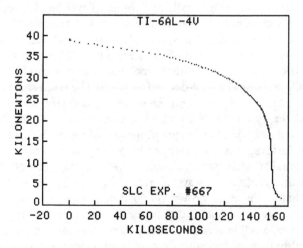

FIG. 3—*Example of load-versus-time data for sustained-load cracking in a Ti-6Al-4V compact tension specimen.*[1]

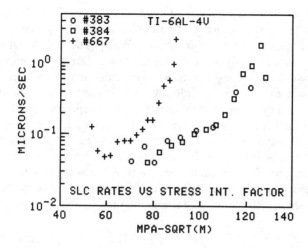

FIG. 4—*Crack growth rate data computed from load-versus-time data. Data for No. 667 obtained using computer system, data for Nos. 383 and 384 obtained using strip chart recorders (see text).*

before acquisition of the computer system, but is included to show the potential of the load drop calibration technique for automated crack growth monitoring.

Figure 5 is an illustration of the speed with which the system can react. The four data points were taken while a specimen was being loaded up. The load-up routine detected a rapid decrease in load, and transferred control to the DATASCAN routine in time to follow the fracture of the specimen. The first

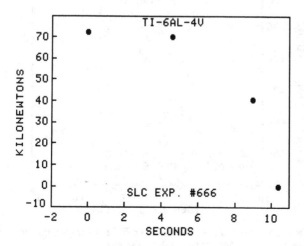

FIG. 5—*Load-versus-time data for sustained-load cracking experiment showing capture of data when specimen broke during initial load-up.*

data point represents the highest load reached during load-up; the other three were taken after control transferred to the DATASCAN routine.

Fatigue—Figure 6 shows an *S-N* fatigue life plot of data obtained using this system. The results clearly demonstrate the loss in fatigue strength incurred by the laser surface melting and rapid resolidification of an approximately 100-μm-deep layer. The method initially used to detect cycles-to-failure relied upon mechanical cycle counters, a spring to lift the specimen load arm upon specimen fracture (to preserve the fracture surface), and a microswitch on the arm to stop the counter and drive motor when the arm sprang up. This method gave erratic results and unusually long lives because of the load shedding characteristic of the displacement-controlled loading system. The use of the computer with load cells to detect the end of useful life gives more uniform results consistent with handbook fatigue lives for similar materials.

Discussion

This system has performed satisfactorily, especially as far as the computer and its plug-in accessories are concerned. All problems resulting in erratic data, noise, changes in calibration, and the like have been traced to defects in connections, ground paths and shielding in analog signal circuits, and to malfunctioning amplifiers in the LCCA. Many of the problems cited above can be expected to dissipate or at least become less serious with the use of the second-generation 16-channel 12-bit ADC now in operation, partly because of great simplification in the analog circuit paths resulting from its use. The stability of the analog signal amplifiers, both in the LCCA's and in the ADC, have

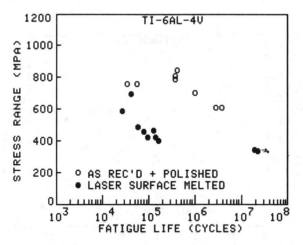

FIG. 6—*Stress versus cycles-to-failure* (S-N) *plot for Ti-6Al-4V cantilever bend fatigue experiments. Arrows indicate test stopped before failure was detected.*

been as specified by the manufacturers. Zero drift and calibration drift together have been much less than 1% overall, making it unnecessary to routinely verify calibration of dial setting. Initially there was a tendency to continually check on progress, but as experience was accumulated (and as the operating program was debugged and made more sophisticated), there developed a feeling that the system was well-trained and reliable.

The conversion of data to binary digital format and its storage in computer-accessible format on magnetic disks have been a major improvement over traditional methods, both as to accuracy and reliability of storage, and as to subsequent processing, analysis, and presentation of the data. The data can be automatically re-processed for storage in other central data storage systems or for further analysis on other data processing systems; it can be printed out, plotted in a variety of formats, transmitted to other facilities, and so on, without significant risk of error and without necessity for manual transcription from paper or display devices. Furthermore, the computer, ADC and disk recording system need not cost much more if at all more than a high-quality strip chart recorder system. The freedom from ink pen problems, mechanical malfunctions, and storage of rolls of recorder paper adds to the advantages of the system. A further advantage of the "intelligent" control of data acquisition is that it is not necessary to examine huge quantities of data such as even a data logger might produce, and that the rate of data acquisition follows precisely the rate of activity of crack growth. The latter characteristic ensures good time resolution during the part of the test where most of the crack growth occurs, which is also the period of most rapid crack growth (or load drop—see Fig. 3). It is difficult or impractical to do this effectively with constant acquisition-rate devices such as strip chart recorders and data loggers. To provide adequate resolution at the rapid period of crack growth, the onset of which is not predictable, the chart speed or printer speed must be set so high that one is inundated with paper or printout tape, and changing paper rolls and replacing ink becomes an onerous task. Furthermore, one may be changing paper while the most significant part of the test is occurring.

It is obvious that a DATASCAN routine written in BASIC would not be suitable for recording dynamic events such as rapid-rate tension tests or to track fatigue waveforms, for example. Such operations are well within the capabilities of the microcomputer and the ADC and clock for tests of moderate speed, if the data acquisition programs are written in assembly code or possibly in FORTRAN. These languages create machine codes which run many times faster than BASIC, which must be interpreted as it runs. Such applications are not presently envisaged because of the difficulty of writing, debugging, and verifying FORTRAN or (especially) assembly code programs. Given the widespread use of microcomputers, it is very likely that such programs exist, and it is to be hoped that they will be publicized so that others can use them.

Concluding Remarks

This paper was written to indicate the impact inexpensive but very capable new devices have made and are making on the automation of tests and data acquisition. It is perhaps unfortunate that so much of what has been done by others was not put to use by the authors (especially operating program software). Perhaps the primary blame for this situation falls on the continued and apparently unstoppable proliferation of ever cheaper, ever more capable and easier-to-adapt computers and interface systems on the market, and on the unbelievably rapid developments in computer operating systems encouraged by cheaper memory systems. These factors make apparently obsolete perfectly good and usable devices and systems within very short times. The system described here could not have been put together five years ago at anywhere near our costs, and at the time it was first developed (about three years ago) there was no software available to do what the authors wanted to do with the computer (Apple II+) of choice. Such rapidity of change makes it difficult to get information to people who might want it in time to avoid apparent obsolescence, and is probably partly responsible for what might be called the Hamming Syndrome (see Ref 7, pages 2 and 3) of people standing on each other's feet instead of on the shoulders of previous workers. One hopes that organizations like ASTM through its committees and by means of publications like this can convince computer equipment manufacturers of the need to standardize operating systems, interconnect hardware, interfaces, and especially computer languages so that programs and system designs can be made truly portable, enabling others to use previously developed techniques on their own equipment.

Acknowledgments

Thanks are due to H. C. Wade for performing the experimental work. This work was supported in part by the Naval Air Systems Command.

References

[1] Joyce, J. A., Hasson, D. F., and Crowe, C. R., "Computer Data Acquisition Monitoring of the Stress Corrosion Cracking of Depleted Uranium Cantilever Bend Specimens," *Journal of Testing and Evaluation*, Vol. 8, No. 6, Nov. 1980, pp. 293–300.

[2] Cullen, W. H., Menke, B. H., Watson, H. E., and Loss, F. J., "A Computerized Data Acquisition System for High Temperature, Pressurized-Water Fatigue Test Facility," *Computer Automation of Materials Testing, ASTM STP 710*, B. C. Wonciewicz, Ed., American Society for Testing and Materials, Philadelphia, 1980, pp. 127–140.

[3] Meyn, D. A., "Effect of Hydrogen Content on Inert Environment Sustained Load Crack Propagation Mechanisms of Ti-6Al-4V," *Environmental Degradation of Engineering Materials in Hydrogen*, M. R. Louthan, Jr., R. P. McNitt, and R. D. Sisson, Jr., Eds., Laboratory for the Study of Environmental Degradation of Materials, Virginia Polytechnic Institute, Blacksburg, VA, 1981, pp. 383–392.

[4] Meyn, D. A., "A Simple Compliance Technique for Estimating Crack Length in Sustained

Load Cracking Tests," *Report of NRL Progress*, Naval Research Laboratory, Washington, DC, Sept. 1977, pp. 4-7.

[5] Yoder, G. R., Cooley, L. A., and Crooker, T. W., "Procedures for Precision Measurement of Fatigue Crack Growth Rates Using Crack-Opening Displacement Techniques," *Fatigue Crack Growth Measurement and Data Analysis, ASTM STP 738*, S. J. Hudak, Jr., and R. J. Bucci, Eds., American Society for Testing and Materials, Philadelphia, 1981, pp. 85-102.

[6] Bayles, R. A., Meyn, D. A., and Moore, P. G., "Laser Processing of Titanium-6Al-4V," *Lasers in Metallurgy*, H. Mukherjee and J. Mazumder, Eds., The Metallurgical Society of the American Institute of Mining, Metallurgical and Petroleum Engineers, 1981, pp. 127-135.

[7] Wonciewicz, B. C., Storm, A. R., and Sieber, J. D., "Systems for Laboratory Automation," *Computer Automation of Materials Testing, ASTM STP 710*, B. C. Wonciewicz, Ed., American Society for Testing and Materials, Philadelphia, 1980, pp. 2-10.

Robert S. Vecchio,[1] David A. Jablonski,[2] B. H. Lee,[2]
R. W. Hertzberg,[3] C. N. Newton,[3] R. Roberts,[3] G. Chen,[3] and
G. Connelly[3]

Development of an Automated Fatigue Crack Propagation Test System

REFERENCE: Vecchio, R. S., Jablonski, D. A., Lee, B. H., Hertzberg, R. W., Newton, C. N., Roberts, R., Chen, G., and Connelly, G., **"Development of an Automated Fatigue Crack Propagation Test System,"** *Automated Test Methods for Fracture and Fatigue Crack Growth, ASTM STP 877*, W. H. Cullen, R. W. Landgraf, L. R. Kaisand, and J. H. Underwood, Eds., American Society for Testing and Materials, Philadelphia, 1985, pp. 44-66.

ABSTRACT: An automated fatigue crack growth rate test system was developed to characterize the crack growth response of engineering materials as a function of the applied stress-intensity range. The test system utilizes DEC PDP 11 series minicomputers and a software feedback control loop to control testing at frequencies up to 100 Hz. Additional system flexibility is provided in its three control modes: constant-load amplitude, constant ΔK, and variable-ΔK test capabilities.

A comparison of crack growth rate data obtained using this automated system is made with data generated under manually controlled test conditions, for aluminum and steel alloys, and for engineering plastics.

KEY WORDS: automation, fatigue crack propagation, computer control, compliance, fracture mechanics

Nomenclature

a Crack length
a_0 Initial crack length
B Specimen thickness
B_{net} Specimen net thickness

[1]Lucius Pitkin, Inc., New York, NY 10013.
[2]Instron Corp., Canton, MA 02021.
[3]Materials Research Center, Lehigh University, Bethlehem, PA 18015.

B_{eff} Effective thickness

C_i Polynomial coefficients used to calculate crack length

C Gradient coefficient in variable K option (length^{-1}) [$C = (1/K) \cdot dK/da$]

COD   Crack opening displacement

da/dN Fatigue crack growth rate

E' Modulus of elasticity

    $E' = E$ plane stress

    $E' = E/(1 - \nu^2)$ plane strain

ΔE_{er} Modulus measurement error

g_f Gage factor for Krak-gages (mm/V)

i Subscript indicating initial value

K Stress-intensity factor

K_{max} Maximum stress-intensity factor

K_{min} Minimum stress-intensity factor

ΔK Stress-intensity factor range ($K_{max} - K_{min}$)

ΔK_{th} Threshold stress-intensity factor range

ΔK_b Baseline stress-intensity factor range

ΔK_{OL} Overload stress-intensity factor range

N Cycle number

N_d Cyclic delay after overload

P Load

P_{max} Maximum load

P_{min} Minimum load

ΔP_{er} Load measurement error

R Stress ratio ($R = P_{min}/P_{max}$)

U Transfer function

U_x Transfer function at position, x, from loading line

U_0 Transfer function at surface of specimen

u Measured voltage potential from Fractomat Krak-gage

u_0 Initial potential from Fractomat Krak-gage

$V_{x/P}$ Inverse slope of load-COD curve where COD is measured at a location x from loading line

V_0/P Inverse slope of load-COD curve where COD is measured at surface

ΔV_{0er} Error in displacement measurement

W specimen width

The concepts of defect tolerant designs and of linear elastic fracture mechanics (LEFM) have made important the measurement of fatigue crack growth rates (FCGR). Since FCGR data may span over seven orders of magnitude from 10^{-10} m/cycle to 10^{-3} m/cycle, their generation involves considerable testing time and expense. Furthermore, different test procedures are used to measure fatigue crack growth rates at the growth rate extremes.

With the advent of computer-controlled fatigue crack propagation testing systems, significant savings in testing time and labor costs have been realized. In addition, the ability to conduct more complicated tests represents, perhaps, the greatest benefit. This latter aspect of computer-controlled testing is particularly important in the development of threshold fatigue crack propagation data associated with extremely low growth rates which require around-the-clock testing. Such computer-controlled threshold testing is usually accomplished under decreasing-K testing conditions through control of the stress-intensity factor gradient, C [where $C = (1/K) \cdot dK/da$]. The advantages of variable-K controlled and constant-K controlled fatigue growth tests have been described elsewhere [1] for the case of compact tension specimens based on compliance-inferred measurement of crack length.

Liaw et al [2] reported a similar technique to measure crack growth rates using a three-point bend specimen with crack length measured by compliance and by an indirect potential method based on a Fractomat.[3] These test results showed excellent agreement between the two crack length measuring techniques and that threshold crack growth rates could be measured using this K-decreasing scheme. An inexpensive digital and analog system to measure near-threshold crack growth rates were developed by Brown and Dowling [3].

This paper examines a computer-controlled material test program, known as FCPRUN, recently developed by the Instron Corp. in cooperation with Lehigh University's Materials Research Center. The essential feature of the program is its ability to control a fatigue crack propagation test using compact tension, wedge opening load, and three-point-bend-type specimens[4] from the threshold regime to unstable fracture under a wide range of driving force conditions (for example, K-controlled, constant-load amplitude, variable-R ratio). In addition, specimens with and without side grooves can be evaluated with this program.

The objectives of this paper are twofold: First, the various features of the computer program FCPRUN are described in detail and FCGR test results are compared using compliance and Fractomat gage techniques. Second, meaningful FCGR data from metals and plastics[5] are compared with valid manual test data derived from constant-amplitude and variable-amplitude loading conditions. To this end, standard plots of crack growth rate, da/dN, versus the crack driving force or stress-intensity factor range, ΔK, are used for data comparison. In addition, a comparison of the cyclic delay behavior of several metallic materials obtained under both manual and computer control is also made.

---

[3] Fractomat and Krak-gage are registered trademarks of TTI Hartrun Corp., Chaska, MN 55318.

[4] The FCPRUN program has recently been modified to accommodate single-edge-notched, center-cracked tension, and round compact tension-type specimen configurations. (For compliance coefficients of the round compact tension specimen, see Ref 4.)

[5] Since the compliance technique is based on linear elastic principles, the viscoelastic nature of certain engineering plastics may introduce fundamental problems when trying to apply this technique.

System Hardware

The test system consists of the following hardware components as shown in Fig. 1:

- servohydraulic test systems,
- computer Interface Unit, and
- minicomputer and associated peripherals.

An Instron Model 1350 servohydraulic test system was used for this investigation. Manual control electronics of the servohydraulic system were Instron Model 2150 controllers which include a hydraulic control module, an analog function generator, digital readout module, load, stroke, and strain controllers, and associated recorders and oscilloscopes. Computer interface electronics consisted of an Instron machine interface unit (MIU), 12-bit analog-to-digital converter (A/D), a 16-bit digital-to-analog converter (D/A) and an 80-bit Parallel I/O (PIO) board. The computer hardware provides control, signal conversion, and analog signal buffering.

To demonstrate the flexibility of the automated test system, PDP 11/03, 11/23, and 11/34 minicomputers (manufactured by Digital Equipment Corp.) were used during this investigation.

System Software

The computer-controlled material testing system was run under the real-time operating system RSX11M (designed by Digital Equipment Corp., for multitasking and multiprogramming environments), with FCPRUN written in FORTRAN to take advantage of the flexibility of a high-level language. In order to have an application program monitor and control the material testing machine, an Instron machine driver (IMD) is used as an interface module. The IMD is a collection of macro-subroutines which handle most of the complex and repetitive tasks of controlling the test machine. The use of the IMD with an application program written in FORTRAN provides a simple method to command machine functions with single call statements. Thus, these FORTRAN callable subroutines in the IMD provide versatile high-level access to the testing machine monitor and control functions.

In order to relate the analog signals of the testing machine to the physical units used in the application program, a servohydraulic machine calibration program (CALSH) was used. This program calculates a set of calibration constants and stores these constants in a disk resident file for later retrieval.

There are three program tasks included in the fatigue propagation software system: FCPINP, FCPRUN, and CRKFIT. The interaction of the various program tasks is illustrated in Fig. 2. FCPINP creates a file containing the test parameters used to specify and control the waveform results, and termination criteria for a test. FCPRUN runs the test, analyzes and reports results to the printer, and writes test results to a permanent disk file, which is avail-

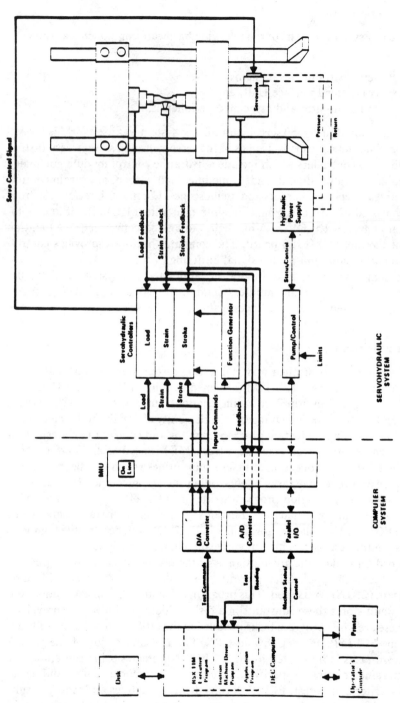

FIG. 1—Schematic diagram of computer-controlled test system.

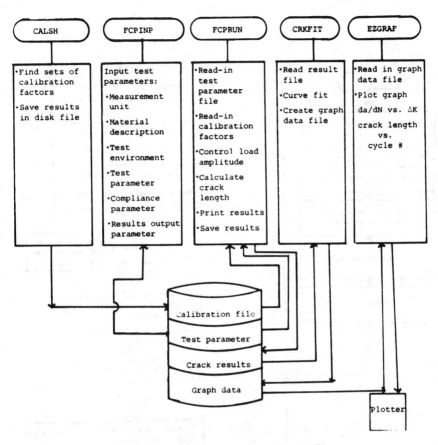

FIG. 2—*Interaction of program tasks for automated FCP test system.*

able for an additional user program level (see Fig. 3 for flow chart of this system). The main functions of FCPRUN are designed to set up the test parameters, start up the servohydraulic machine, monitor and control the machine, analyze the data, generate the report of results, and provide for on-line parameter changes. The first phase of this program is to install the test parameters previously created by using FCPINP. Then, the following list of additional parameters is entered in FCPRUN before actually starting the test: servohydraulic machine number, machine control mode, maximum load and load ratio (R), current crack length, and final crack length for terminating test.

Under K-controlled test conditions the maximum load value is used as a software load limit. Otherwise, maximum load is the load to be controlled. The program will stop the test if the command load for the given value of DELTA K to be controlled is greater than this maximum load setting. Next, the program prints messages to guide the operator through the installation of

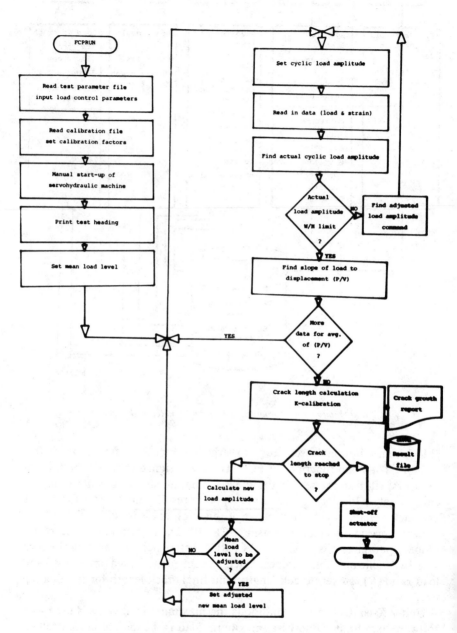

FIG. 3—*Flow diagram for test control program (FCPRUN).*

the specimen and manual starting of the machine. When the operator places the machine on line, the computer has total control. The program then reads in the calibration constants stored in the calibration file, and a test report heading is printed on the terminal printer.

The first machine control command is to bring the specimen to the mean load level. A cyclic load is then expanded about the mean load level in successive increments of 60%, 20%, and 20% of maximum load level. This scheme was adopted to prevent overshooting of load when the analog function generator is activated. The adjustment of the load amplitude command is done in a manner to reduce the difference observed between the desired load and actual feedback reading of the load channel. In FCPRUN, a software gain is introduced in such a way that load amplitude command to the testing machine is the product of a software loop gain and the desired load amplitude. Correction to the load command is made every two seconds. Some of the test parameters can be changed while the test is in progress. During this mode of operation, the operator has the option of either continuing cyclic loading or stopping the cyclic load at the current mean load level.

FCPRUN continuously monitors specimen crack length by using the elastic compliance technique or the Fractomat. Analysis of data begins by reading 400 data points from three channels (load, strain, and auxiliary). The 400 data points are analyzed to calculate the compliance, peak-to-peak load amplitude of the waveform, and the crack length from the Fractomat. Minimum and maximum data values from each cycle are found and then averaged to calculate the best estimate of actual peak-to-peak values of test wave shape. Actual mean load level is determined from these averaged minimum and maximum values.

The number of data pairs available for calculating compliance is a function of test frequency, with a minimum of about 20 pairs for each half of the loading cycle. The slope of load-displacement curve that is required in the compliance calculation is found by linear regression analysis of these data pairs for each loading-unloading cycle. The operator has control over which data pairs are used for slope determination, including selection of intermediate points of the load-displacement curve, such that only the linear section of the curve is used; selection of either the loading, unloading, or both portions of the load-displacement curves; and selection of the number of multiple sets of measurements to determine the slope. In order to accommodate high test frequencies, the program used a burst mode data acquisition method. In burst mode, data collection is on two feedback channels and the interval timer on the A/D board is programmed differently, thereby allowing data to be collected at a higher rate than in the normal data collection mode. The burst mode data acquisition allows a range of data rates from 250 μs to 1 ms. In the case of a high-frequency test, however, the response of the testing system components, such as a servohydraulic actuator and the displacement transducer, tends to become the limiting factor.

Finally, test results are printed on the terminal printer when the crack propagates a user-specified crack increment. The results are also saved on a disk file for later reference. FCGR are calculated using the modified secant method, which represents the change in FCGR between the previous and next crack length report.

CRKFIT analyzes the result file data according to the seven-point polynomial method specified in Appendix XXI of the ASTM Test Method for Constant-Load-Amplitude Fatigue Crack Growth Rates Above 10^{-8} m/Cycle (E 647-83), and creates data files that are compatible for plotting.

Experimental Test Procedures

Materials used in this study included two 2024-T351 aluminum alloys, A514F, 1035, and A588 steels, polycarbonate, polystyrene, and epoxy resins. Mechanical properties are summarized in Table 1. Three specimen geometries were used in this study: ASTM standard compact tension, three-point bend, and wedge opening loaded (WOL) type specimens.

Crack length was determined by using either unloading compliance or Fractomat gages. In the compliance method the linear region of the load-versus-COD curve which was used to calculate the slope is defined in terms of a percentage of the load range and was a user-defined option to account for crack closure. To calculate the crack length, the program first determines, by least-squares analysis, the slope of the load-versus-COD curve between the specified limits which is then input into the following equations to calculate the crack length [5–7].

$$a/w = C_0 + C_1 U_x^1 + C_2 U_x^2 + C_3 U_x^3 + C_4 U_x^4 + C_5 U_x^5 \qquad (1)$$

TABLE 1—*Material properties.*

A: Metals Material	σ_y, MPa	σ_u, MPa
2024-T351	325 to 362	450 to 496
A514	690	890
A588	345	485
1035	517	586

B: Plastics Material	σ_y, MPa	MW
PC	63	37 000
PS	40	450 000
Epoxy	55	. . .

where U_x is a transfer function defined as follows:

$$U_x = 1 \Big/ \left[\left(\frac{E' V_x B_{eff}}{P} \right)^{1/2} + 1 \right]$$

$$B_{eff} = B - (B_{net})^2/B$$

(2)

Crack lengths were also measured with a Krak-gage, a thin metallic foil on a polymeric backing which is adhesively bonded to the specimen. The Fractomat 1078 instrument applies a constant current to the gage and the instrument measures potential u. Such gages are designed so that the potential Δu is directly proportional to the crack length. The crack length is calculated by

$$a = a_0 + (u - u_0) \cdot g_f$$

(3)

The voltage output from the Fractomat varies during dynamic cycling because of periodic shorting of the crack gage surfaces at minimum load. Periodic variations of voltage were eliminated by use of the peak reading circuit, which measures only the peak voltages during cycling, thus maintaining the voltage output as continuous. Since the Fractomat is a two-channel device which allows the crack length to be measured on both surfaces, it is possible to obtain an average crack length reading.

The stress-intensity factor was calculated by use of standard equations from ASTM E 399. For the case of side-groove specimens, the stress-intensity factor was calculated by

$$K_{side\ grooves} = \sqrt{\frac{B}{B_{net}}} \cdot K_{no\ side\ grooves}$$

(4)

Crack growth rates during the run time section of the test were calculated by the modified secant method

$$(da/dN)_i = \frac{a_{i+1} - a_{i-1}}{N_{i+1} - N_{i-1}}$$

(5)

For some of the data reported, crack growth rates were fit by a seven-point incremental polynomial.

Both ΔK increasing and decreasing test conditions were controlled at the user's discretion based on

$$\Delta K = K_i \exp C(a - a_i)$$

(6)

where $C = (dK/da)/K$ (the stress-intensity factor gradient). When C is less than zero, a K-decreasing test is specified.

In all manually controlled fatigue tests, crack extension was monitored using a Gaertner traveling microscope, typically at growth intervals of 0.2 to 0.25 mm. Load cycling was interrupted to make these measurements and to record the associated number of loading cycles (N). Constant baseline stress-intensity conditions (ΔK_b) were achieved by shedding loads (less than 1% increments) at least every 0.2 mm. Delay behavior was determined by applying single peak tensile overloads at a frequency equal to or less than 0.2 Hz (see Ref 8). The percentage overload (% OL $= \Delta K_{OL}/\Delta K_b \times 100$) was kept constant at 100%. The amount of cyclic delay, N_d, due to an overload was measured from the crack length (a) versus cycles (N) curves (see Ref 8).

Results and Discussion

The experimental results section is divided into two main parts. First, representative results are shown for 2024-T351 to demonstrate the application of FCPRUN and the use of both crack measuring techniques. Second, a comparison of computer versus manually generated FCGR information is presented and discussed.

The results of a constant-ΔK tests are shown in Fig. 4 and reveal that crack growth rates are independent of crack length for both compact tension and

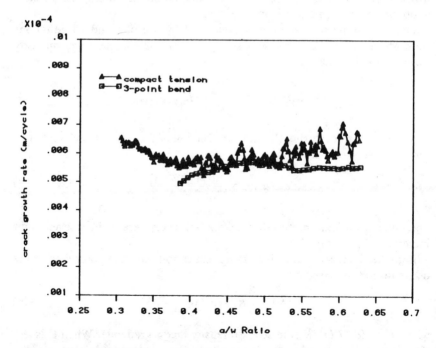

FIG. 4—*Variation in FCGR as a function of* a/w *for compact tension and three-point bend specimen (*$\Delta K = 16.5\ MPa\sqrt{m}$, R $= 0.1$*).*

three-point bend specimens, as expected. The maximum variation of crack growth rate from the mean is only 13%. The greater amount of scatter in the compact tension specimen data is believed to be due to the larger amount of error in the crack length measurements; reasons for this are discussed in the Appendix. Figure 5 shows an FCGR curve for a three-point bend specimen tested under load control at $R = 0.50$, based on crack growth rates measured by both compliance and Fractomat methods; excellent agreement is seen.

In K-decreasing tests, crack closure usually develops in a near-threshold crack growth rate regime. In addition, crack closure forces are observed to increase as the threshold is approached [9–11]. An example of increasing crack closure is presented in Fig. 6 for a K-decreasing test conducted at $R = 0.1$. Note how the load-versus-COD curve becomes increasingly nonlinear as ΔK decreases. An estimation of the ratio of the opening load to the maximum load is made by determining the point on the upper section of the curve where the deviation from linearity occurs. Figure 6c reveals that near the measured threshold, crack closure appears to influence the entire load-COD curve, thus making it impossible to find a linear section to calculate crack length. The significance of this observation is discussed at greater length later.

This section of the paper now focuses on an evaluation of the suitability of using FCPRUN to monitor FCGR response in numerous ferrous and nonferrous alloys, and engineering plastics.

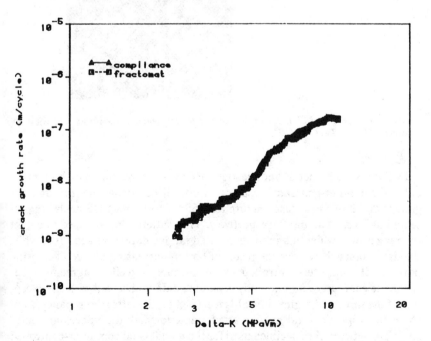

FIG. 5—*Comparison of FCGR measured by compliance and Fractomat techniques in three-point bend specimen (R = 0.5).*

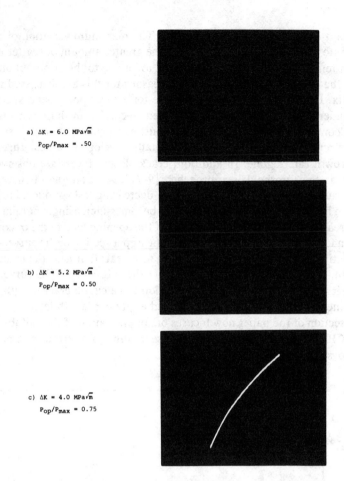

a) $\Delta K = 6.0$ MPa$\sqrt{m}$
$P_{op}/P_{max} = .50$

b) $\Delta K = 5.2$ MPa$\sqrt{m}$
$P_{op}/P_{max} = 0.50$

c) $\Delta K = 4.0$ MPa$\sqrt{m}$
$P_{op}/P_{max} = 0.75$

FIG. 6—*Photos of load-COD curves showing the development of crack closure as* ΔK_{th} *is approached (A1 2024-T351,* R $= 0.1$*).*

FCGR results in metals and plastics are now compared based on both manual and computer-generated test results. FCGR data from the threshold regime to ΔK levels near fast fracture for A588, A514, and 1035 steels are presented in Figs. 7, 8, and 9, respectively. The scatterbands are used to depict the ranges over which manually generated fatigue data were acquired, while the data points shown are the result of both K-increasing and K-decreasing portions of computer-controlled test conditions. Excellent agreement between the computer and manually generated data is observed with the exception of the threshold regime in A514 steel (Fig. 8); the latter discrepancy is not of major proportion and may be attributable to small differences in microstructure between the specimen used for the manual and computer-controlled tests. (It should be noted that variations in microstructure such as grain size

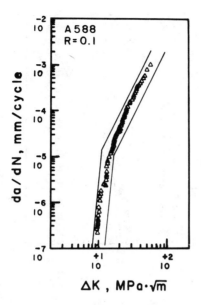

FIG. 7—*FCP response of A588 at an R ratio of 0.1. The scatterband depicts data obtained under manual control, while the symbols are data-generated under computer control.*

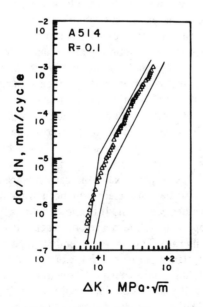

FIG. 8—*FCP response of A514 at an R ratio of 0.1. The scatterband depicts data obtained under manual control, while the symbols are data-generated under computer control.*

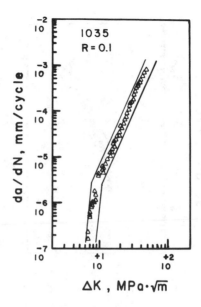

FIG. 9—*FCP response of 1035 at an* R *ratio of 0.1. The scatterband depicts data obtained under manual control, while the symbols are data-generated under computer control.*

do have a significant influence on threshold behavior [*12,13*].) Although the difference between the computer and manually generated data for A514 (Fig. 8) is small, it has been the experience in this laboratory that considerable discrepancy in the threshold regime can occur if inference of the crack length is in error. This results from the fact that small changes in the stress-intensity range (due to errors in crack length inference) in the threshold regime result in substantial changes in the corresponding crack growth rate. Though the data shown in Fig. 8 reveal a slightly conservative estimate of the ΔK_{th} relative to the manually determined value, it should be noted that errors in the inferred crack length generally lead to an overestimation of a material's threshold characteristics. Such overestimations of ΔK_{th} have been referred to as "false" thresholds.

If one considers a role of crack closure or crack surface interference on the near-threshold regime, it becomes clear how both crack arrest and overestimation of threshold values can occur. It is known that as the threshold stress-intensity range is approached, the degree of crack surface interference increases at an increasing rate. This, in turn, leads to an ever-decreasing linear portion of the load-displacement traces. For example, if the operator uses the compliance technique and chooses a point on the load-displacement curve (Fig. 6) which is below the crack opening level, then the computer will monitor highly nonlinear load displacement traces. As a result, the computer will infer the crack length to be shorter than its actual size. Therefore, when conducting tests under computer control in the near-threshold regime, the operator should carefully monitor the shape of the load-displacement curve and periodically verify by visual reading the actual crack length.

The fatigue crack propagation response of polycarbonate and polystyrene was examined under both manual and computer-control test conditions (Figs. 10 and 11). Again, as was the case with the metals, these amorphous polymers exhibit excellent agreement between the computer and manually generated test data. The ability to conduct reliable K-increasing and K-decreasing tests in these materials is also illustrated in Fig. 10 where such data are seen to compare well with the manually controlled constant-load amplitude results.

It is known from manual fatigue crack growth rate test results that polystyrene exhibits a modest sensitivity to the frequency of the applied loads (Fig. 11). Since the lower-growth-rate computer-controlled data are in excellent agreement with the higher-growth-rate manually generated data for two different frequencies (10 Hz, Fig. 11, and 50 Hz, Fig. 11), it is apparent that use of the compliance technique in this computer-controlled application is sensitive enough to detect these frequency-induced variations in crack growth rate. Other fatigue tests of an epoxy resin which is not frequency sensitive further verify the suitability of using FCPRUN to monitor fatigue crack growth in glassy, amorphous plastics (Fig. 12).

The use of this program, based on compliance-based inference of the crack length, is brought into question, however, when dealing with more compliant

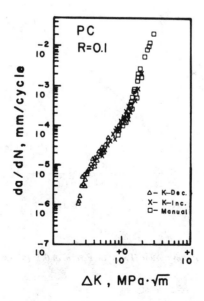

FIG. 10—*FCP response of polycarbonate (PC) at an R ratio of 0.1.*

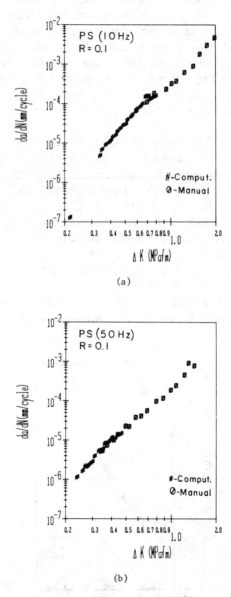

(a)

(b)

FIG. 11—*FCP response of polystyrene (PS) at an R ratio of 0.1: (a) test frequency of 10 Hz; (b) test frequency of 50 Hz.*

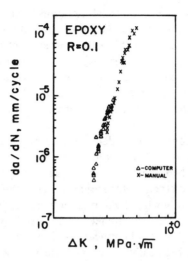

FIG. 12—*FCP response of epoxy resin at an R ratio of 0.1. Test frequency of 40 Hz.*

polymers—especially those that experience a considerable amount of hysteretic heating. In the latter instance, such heating often leads to a decrease in the stiffness of the unbroken ligament of the specimen. It follows that if the material's modulus changes continually during crack extension, crack length inferences via the compliance technique will lead to spurious results. In addition, materials that demonstrate considerable viscoelastic response tend to exhibit a load-deflection trace that is generally nonlinear. As such, it is not clear how one may define the slope of this nonlinear plot to infer the current crack length. Attempts to perform a computer-controlled test with a rubber-modified Nylon 66 plastic proved unsuccessful for the above two reasons. It should be pointed out that the difficulty in conducting a computer-controlled test of this material was not due to limitations in the FCPRUN program, *per se*, but rather due to the compliance technique used to infer the crack length.

Variable-amplitude loading can have a pronounced effect on the total fatigue life of a component. For example, it has been shown that single or multiple tensile overloads or compressive underloads can lead to a temporary attenuation or acceleration, respectively, of the FCGR [8,14,15]. The baseline stress-intensity factor range (ΔK_b) has been identified as a major variable that influences a component's response to overloads. As such, variable crack growth information can be obtained from a particular material by conducting overload studies at various constant baseline stress-intensity ranges from the threshold regime to those approaching fast fracture. In order to more clearly identify the role of the baseline stress-intensity range in variable-amplitude loading it is useful to conduct these tests under constant-ΔK test conditions.

In this investigation, several 100% tensile overload computer-controlled experiments were conducted on 2024-T3 aluminum and A514F steel. The results of these tests and the results from identical but manually controlled experiments are presented in Table 2. Examination of these data reveals that the computer-controlled results consistently underestimated the amount of cyclic delay due to a tensile overload as determined separately from manual test procedures. Though there is some scatter observed when conducting manual overload tests, it is not sufficient to explain the large differences shown in Table 2.

The underestimation of cyclic delay due to an overload using computer control is believed to be directly related to an error in the crack length inference through use of the compliance technique. It is known that tensile overloads result in residual crack surface interference in the wake of an advancing crack tip. Therefore, following the application of a tensile overload, a considerable increase in the crack closure level is observed. As the crack tip moves away from the point of overload application, the closure level first increases to a maximum and then gradually decreases to the pre-overload level. Figure 13 illustrates the shape of two typical load-displacement traces; trace (1) is a

TABLE 2—Cyclic delay data.

Material	ΔK_b	Manual Control Cycles of Delay $\times 10^3$	Computer Control Cycles of Delay $\times 10^3$
2024-T3	7.5	100 to 125	40 to 60
A514	12.5	75 to 80	50
A514	10.0	300	70 to 110

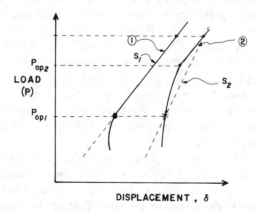

FIG. 13—*Schematic illustration of load-displacement curve prior to overload [Trace (1)] and load-displacement curve following a tensile overload [Trace (2)]. Erroneous estimate of slope S_2 results from use of incorrect value of P_{op}. Note dramatic increase on crack opening level following overload [Trace (2)].*

load-displacement plot which is observed just prior to a tensile overload and trace (2) is typical of a load-displacement plot obtained following the overload. In Fig. 13, P_{op1} and P_{op2} are the crack opening levels associated with Traces (1) and (2), respectively. It is obvious from this figure that Trace (1) has a low P_{op1} level which results in a substantial linear load-displacement region; this in turn yields an easily measured slope, S_1. Trace (2), on the other hand, has a very high P_{op2} level which results in a much reduced linear region. Assuming that the opening levels are the same in the two curves, the slope S_2 of Trace (2) is therefore approximated by what amounts to a secant to this curve. Clearly, S_2 is greater than S_1. As such, the inferred crack length a_2 is less than a_1, and the resultant applied stress σ_2 is greater than the previous value σ_1. Therefore, the amount of cyclic delay under computer control will be less conservative when compared with manual test data because of the higher applied stress range.

Conclusions

1. A considerable savings in time and money is realized when standard fatigue crack propagation tests are conducted under computer control.

2. It was shown that the automated FCGR program was capable of measuring crack growth rates over five orders of magnitude.

3. Operators of computer-assisted facilities should be aware of the influence of crack closure on the linearity of the load-COD trace used to infer crack length when conducting decreasing-K threshold tests.

4. The factors which affect the crack length accuracy are errors in the load and COD measurement, errors in the modulus, and the percentage of the load-COD signal used for the slope determination.

5. Agreement between computer and manually generated data is excellent for metals and glassy, amorphous polymers.

6. Highly viscous materials such as Nylon 66 are not presently amenable to compliance-based computer control, since testing of these materials results in very nonlinear load-displacement traces.

7. The use of the compliance technique for inferring crack length in variable-amplitude loading studies yielded consistently more conservative cyclic delay results than those generated under manual control.

Acknowledgments

The authors wish to express their appreciation to the Instron Corp. and the Materials Research Center, Lehigh University, for providing the support and use of the facilities for this study. Partial support from the U.S. Air Force under Grant No. AFOSR-83-0029 is also recognized. The authors would also like to acknowledge Ms. J. Prosperi for developing burst mode programming, Mr. M. Ritter for coding the Fractomat section of the FCP program, and Dr.

J. Lin and Mr. P. Bensussan for their many helpful technical discussions. Thanks are extended to R. Lang, D. Rohr, and J. Michel for providing a portion of the manual test data. Finally, the assistance of Louise Valkenburg, Betty Zdinak, and P. Papagno for preparing this manuscript is also greatly appreciated.

APPENDIX I

The accuracy of the crack length determination via elastic compliance depends upon many variables. Friction in the load train has been found to affect crack length inference; by reducing pin friction through the use of a needle bearing, crack length sensitivity could be increased from 0.05 to 0.03 mm [16]. Other factors which affect crack length accuracy include errors in load measurement, errors in COD measurement, and errors in modulus information. Therefore, the total error in crack length can be estimated by forming the following partial differential of Eq 1

$$\Delta a_{\text{error}} = W\left(\frac{\partial a}{\partial U}\right)\sqrt{\left[\left(\frac{\partial U}{\partial E}\,\Delta E_{\text{er}}\right)^2 + \left(\frac{\partial U}{\partial P}\,\Delta P_{\text{er}}\right)^2 + \left(\frac{\partial U}{\partial V}\,\Delta V_{0\text{er}}\right)^2\right]} \quad (7)$$

where U is the transfer function defined by Eq 2 in the text.

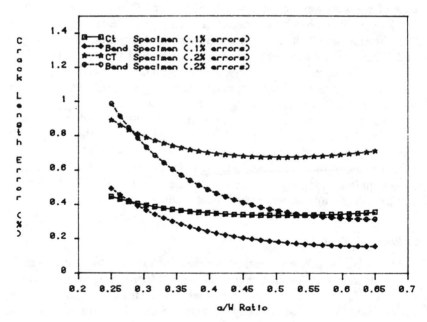

FIG. 14—*Crack error for constant-*K *test with* $\Delta K = 16.5\ MPa\sqrt{m}$.

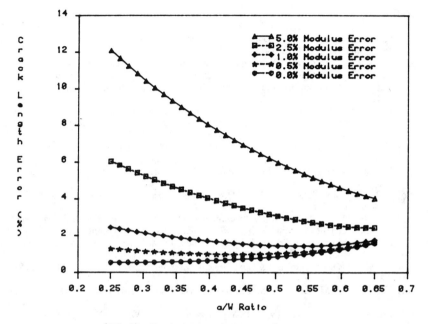

FIG. 15—*Crack error for various modulus errors.*

Performing the appropriate partial differentiation and defining the variable z as

$$Z = \left(\frac{E^1 V_0 B_{\text{eff}}}{P}\right)^{1/2} \left[1\left(+ \frac{E^1 V_0 B_{\text{eff}}}{P}\right]^{1/2}\right]^2 \tag{8}$$

Equation 7 may be reduced to the following expression

$$a_{\text{error}} = \frac{w \cdot z}{2}[C_1 + 2C_2 U + 3C_3 U^2 + 4C_4 U^3 + 5C_5 U^4]$$

$$\times \left[\left(\frac{\Delta E_{\text{er}}}{E}\right)^2 + \left(\frac{\Delta V_{0\text{er}}}{V_0}\right)^2 + \left(\frac{\Delta P_{\text{er}}}{P}\right)^2\right]^{1/2} \tag{9}$$

Figure 14 reveals a plot of percentage crack length error in compact tension (CT) and three-point bend specimens as a function of crack length-to-width ratio for constant-ΔK testing. It is seen that the crack length error decreases as the crack propagates since the increase in specimen compliance is greater than the decrease in load during a constant K test. The effect of an error in elastic modulus on determination of compliance-inferred crack length is shown in Fig. 15. As expected, small moduli errors correspond to small errors in inferred crack length. It is interesting to note that the error in crack length for a large modulus error (5%) decreases with increasing crack length.

References

[1] Saxena, A., Hudak, S. J., Donald, J. K., and Schmidt, D. W., *Journal of Testing and Evaluation*, Vol. 6, No. 3, 1978, p. 167.

[2] Fabis, T. R. and Liaw, P. K., "Computer-Control Fatigue Crack Growth Rate Testing on Bend Bars in Corrosive Environments," Presented at Fall Meeting of the Metallurgical Society of the American Institute of Mining, Metallurgical and Petroleum Engineers, Oct. 25, 1982.

[3] Brown, R. C. and Dowling, N. E. in *Fatigue Crack Growth Measurement and Data Analysis, ASTM STP 738*, American Society for Testing and Materials, Philadelphia, 1981, pp. 58-67.

[4] Jablonski, D. and Vecchio, R. S., "Compliance Coefficients for the 3-Pt. Bend and Round Compact Tension Specimens," manuscript in preparation.

[5] Saxena, A. and Hudak, S. J., *International Journal of Fracture*, Vol. 14, 1978, pp. 453-468.

[6] Shih, C. T. and deLorenzi, H. C., *International Journal of Fracture*, Vol. 13, 1978, pp. 544-548.

[7] Jablonski, D. A. and Lee, B. H., "Automated Fatigue Crack Growth Rate Testing Using a Computerized Test System" in *Proceedings*, SECO '83, Digital Techniques in Fatigue, Society of Environmental Engineers Fatigue Group, London, March 1983, pp. 291-308.

[8] Vecchio, R. S., Master's Thesis, Lehigh University, Bethlehem, PA, 1983.

[9] Ritchie, R. O. and Suresh, S., *Metallurgical Transactions*, Vol. 13A, May 1982, pp. 937-940.

[10] Vasudevan, A. K. and Suresh, S., *Metallurgical Transactions*, Vol. 13A, Dec. 1982, pp. 2271-2288.

[11] Minakawa, K. and McEvily, A. J., *Scripta Metallurgica*, Vol. 15, 1981, pp. 633-636.

[12] Yoder, G. R., Cooley, L. A., and Crooker, T. W., Naval Research Laboratory Memorandum Report 4576, Washington, DC, July 1982.

[13] Yoder, G. R., Cooley, L. A., and Crooker, T. W., *Scripta Metallurgica*, Vol. 16, 1982, p. 201.

[14] Vroman, G., North American Rockwell, Los Angeles Division, T.R. TFR-71-701, 1971.

[15] Schijve, J., *The Accumulation of Fatigue Damage in Aircraft Materials and Structures*, AGARDograph 157, Neuille-sur-Seine, France, 1972.

[16] Hewitt, R., *Journal of Testing and Evaluation*, Vol. 11, No. 2, March 1983, pp. 150-155.

William R. Catlin,[1] *David C. Lord,*[1] *Thomas A. Prater,*[1] *and Louis F. Coffin*[1]

The Reversing D-C Electrical Potential Method

REFERENCE: Catlin, W. R., Lord, D. C., Prater, T. A., and Coffin, L. F., "**The Reversing D-C Electrical Potential Method,**" *Automated Test Methods for Fracture and Fatigue Crack Growth, ASTM STP 877,* W. H. Cullen, R. W. Landgraf, L. R. Kaisand, and J. H. Underwood, Eds., American Society for Testing and Materials, Philadelphia, 1985, pp. 67–85.

ABSTRACT: An automated reversing d-c potential system capable of measuring crack growth rates in compact-type, edge-notched, and surface-defected specimens under a wide variety of environmental conditions is described. Measurements can be made with ease during cyclic or static loading conditions. The unique characteristics of the system software, which give the operator a wide choice of test routines, including constant load range, constant stress-intensity range, and programmed load range or programmed stress-intensity range, are discussed. Detailed procedures are then given for conducting a "controlled K" cyclic test on a surface-defected specimen.

The system has outstanding stability and sensitivity. The high frequency of measurement permits averaging of many readings within a given cycle to obtain a single data point or to detail a given cycle. Techniques for averaging data to enhance the resolution of growth rate measurements are discussed.

Finally, results from several tests are described to demonstrate the versatility of the system in making measurements on different specimen geometries over a wide range of testing conditions.

KEY WORDS: automated fatigue testing, compact tension, surface defect, crack growth, crack monitoring, data acquisition, d-c potential measurement

Although crack growth studies have been made over the years using a variety of monitoring systems, including visual, compliance, and potential-drop methods [1], a need has developed for a method with high resolution and long-term stability that is capable of functioning in environmental chambers under both static and cyclic loading conditions. The reversing d-c electrical potential method has been developed at General Electric's Research and De-

---

[1]Specialist, specialist, metallurgist, and mechanical engineer, respectively, General Electric Co., Corporate Research and Development, Schenectady, NY 12301.

velopment Center to monitor crack growth in test specimens subjected to simulated nuclear power plant environments under these conditions. Historically, drift due to thermal electromotive forces (emf's) created at junction points has caused a problem in making d-c potential measurements. In the present approach, the current is reversed at 0.5 s intervals to minimize this problem. The resulting differential potential readings between any two adjacent readings with positive and negative polarity eliminates concern about thermal emf's and amplifier zero drift if the time between adjacent readings is relatively short.

A variety of test specimen geometries has been employed with this approach, including some which simulate the conditions encountered in service components much more closely than the widely used compact-type (CT) fracture mechanics specimen [2–7].

The basis of the system is to pass a known direct current through a test specimen and to monitor the change in potential drop between one or more probe pairs, referred to as active probes, which span a known defect or crack in the gage section as growth occurs. These probes will react to changes in defect size, temperature, current, and strain deformations. A reference probe pair is located at a location remote to the defect but, where possible, in the specimen's gage section. This probe will react in the same manner to all variables except defect size. All probe pair potential readings are normalized by dividing the instantaneous value for the potential by the potential reading at the same location at the start of testing. A "net ratio" for each active probe pair is obtained by dividing the normalized value of the potential from an active probe by the normalized value of the reference probe. Use of the net ratio for each active probe pair compensates for changes in current, temperature, and strain. Analytical solutions are then used to convert the recorded changes in potential to changes in defect size. Recent advances in electronics and microcomputers have been introduced to allow the reversing d-c electrical potential method to become a very stable, high-resolution system for crack measurement.

System Description

The reversing d-c electrical potential method can be viewed as four systems (see Fig. 1): (1) current control, (2) potential measurement, (3) data acquisition, and (4) test machine control. The microcomputer and its software package are the key to the control of all four systems.

The current control circuit, Fig. 1, includes a d-c power supply, shunt, and solid-state switches. The d-c power supply is capable of current control mode with 0.02% regulation and 0.03% drift characteristics. While amperage varies with specimen material and cross-sectional area, 5 A is more than adequate for all specimens and 50 A for components tested to date. The current shunt is located in the positive lead of the d-c power supply and yields 10 mV

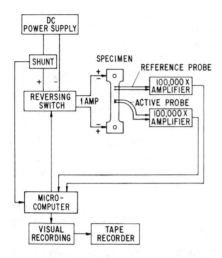

FIG. 1—*Block diagram of components used for microcomputer controlled reversing d-c potential crack growth measurement system.*

analog signal per ampere. The analog signal does not require signal conditioning to the microcomputer and is recorded as a permanent record of the current history of each test. The solid-state switching is located between the shunt and specimen; hence it does not interrupt the shunt's analog signal to the microcomputer. It is controlled by solid-state logic levels of 0/5 V d-c from the microcomputer with a switching rate of $\frac{1}{2}$ s at each polarity throughout the test.

The potential circuit, Fig. 1, includes ×1000 gain isolation amplifiers with high 166 dB common-mode rejection and low long-term drift, typically 2 µV per year. Short, shielded, twisted-pair input leads are used from the platinum entry lead terminal to the amplifiers to minimize electrical noise pickup. Each test facility requires one amplifier per probe pair with outputs at millivolt levels. The signals are inputs to the microcomputer and are first multiplexed to a ×100 gain internal instrument amplifier, sampled, held, and finally digitized. A gain of ×100 000 gain is required because potentials are kept in the microvolt range to minimize electrochemical influence on crack growth. Potential readings are made at each probe position in sequence after current of a given polarity has been flowing for $\frac{1}{2}$ s, 16 potential readings are taken at each probe pair within a few milliseconds. The current polarity is then reversed and after a $\frac{1}{2}$ s interval, another 16 readings are taken at each probe position. One-half the difference between the averages of these readings is used as a single potential reading for that probe position. The $\frac{1}{2}$ s current reversal period is determined by the settling time of the ×1000 gain isolation amplifiers

Data acquisition includes a hard copy printout and storage on magnetic tape. The format varies with test specimen geometry, test conditions, and static versus cyclic loading. An example of dialog and headings is shown in Fig. 2. This format is for cyclic testing of a surface-defected specimen with six active probes and represents the largest quantity of displayed data [6, 7]. The least quantity of data is presented in a single active probe pair static test, Fig. 3. Common to all data outputs are cycle or time count, current, microvolts of all probe pair channels, normalized value of the reference potential, and net ratio of each active probe pair. Additional real-time control outputs include command and feedback minimum and maximum load signals, calculated crack length, and calculated stress-intensity values at minimum and maximum loads. Data are given in scientific notation to reduce the number of characters per line of data output with the mantissa expressed in the heading of each column of data. The dialog varies greatly with specimen geometry, type of test (cyclic versus static) and for passive versus real-time test control.

Test machine control from the microcomputer has been designed to be adaptable to any servohydraulic test machine using a function generator to create a command signal. Incorporation of control requires relocating the function generator's output to the microcomputer and inserting the composite command output of the microcomputer in its place. The mean level and amplitude dials of the test machine are then standardized prior to transferring control to the microcomputer. The function generator must run in a half-waveform mode, that is, zero to +10 V range.

Two digital-to-analog multiplying conditioners (DAC) are dedicated to machine control in the microcomputer. The function of the first is to attenuate the output of a stable +10 V power supply to equal the voltage needed to represent the desired minimum load. It is acceptable to use the mean level dial of the test machine to produce a desired tare load. The dialog will query that value and the first DAC's output will be modified accordingly. The second DAC attenuates the function generator's output to produce the desired amplitude as dictated by the load range specified at that point in testing. The output of both DAC's is summed to produce a composite command output signal that replaces the function generator signal. The amplitude dial of the test machine must be set at full range to avoid attenuation of the microcomputer signal. The microcomputer must be informed of several parameters prior to test initiation; these are load range, tare load, crack depth, and desired minimum and maximum stress intensity.

Control characters are provided to start or terminate testing and to control the format of data. In cyclic testing, data may be printed for each cycle, or, if desired, as averages of data from a block of cycles. Static load testing control characters start and stop an internal clock which maintains an hour count in the data output.

A timeout device is incorporated at the printer terminal and is wired to stop the function generator at zero volt output should the printer not print in a time interval somewhat greater than the anticipated interval of printout. This

RUN
REV DC POT K CNTRL-CYCLIC-DOG BONE-040183-1200

MACHINE
LOAD RANGE (#)? 10000
TARE LOAD (#)? 500

SAMPLE
WIDTH & THICKNESS (".")? .7,.2
ORIGINAL CRACK DEPTH (")? .150
PRESENT CRACK DEPTH (")? .150
ASPECT RATIO (.3=<(N)<=.5)? .5
PROBE SPACING (")? .051

TEST
PRESTART/RESTART CYCLE COUNT (N)? 0
CYCLES FOR INITIALIZATION (N)? 0
WAVE PERIOD (WP=>4) (SEC)? 600
(S)INE OR (T)RIANGULAR? T
DATA THRESHOLD (0=<(N)<=1)? 0

BLK SZ FOR 50 18-PR SCANS = 1

BLOCK SIZE (N)? 1
NUMBER OF BLOCKS PER PRINT-OUT (N)? 1

PROGRAM
CONSTANT MODE TEST (Y)ES OR (N)O? Y
CONSTANT (L)OAD OR (K)? K
K(MAX)? (KSI)? 18
K(MIN)? (KSI)? 1.8
'A' VALUE AT TERMINATION? .155

(CTL/R RUNS F6 & PRINTS CYCLE DATA DETAIL)
(SPACE BAR SETS BLOCK AVERAGE DATA ONLY)
(CTL/I INSERTS NEW CYCLE INFORMATION INTO RUNNING PROGRAM)
(CTL/H HALTS TEST & FUNCTION GENERATOR)

CYC CH 7 E+0	CNT AMPS E-3	REF CH 0 AVG E-8	NORM E-4	ACT CH 1 AVG E-8	NET E-4	ACT CH 2 AVG E-8	NET E-4	ACT CH 3 AVG E-8	NET E-4	ACT CH 4 AVG E-8	NET E-4	ACT CH 5 AVG E-8	NET E-4	ACT CH 6 AVG E-8	NET E-4	CRK LNGTH E-4	K MX CALC E+2	K MN CALC E+2	P MX CMD E+0	P MN CMD E+0	P FB MAX E+0	CH 0A MIN E+0	CH 2A TEMP-F E+0
0	0	0	4205	0	10000	0	10000	0	10000	0	10000	0	10000	0	10000	1500	180	0	4619	462	0	0	0
2	3017	4207	10003	2978	10000	2977	10000	3752	10000	3752	10000	2984	10000	2984	10000	1500	180	18	4619	462	4620	455	78
3	3017	4208	10002	2980	10003	2979	10003	3753	9998	3753	9998	2986	10002	2986	10002	1500	180	18	4618	462	4625	452	77
4	3017	4207	10005	2980	10004	2980	10004	3754	10003	3754	10003	2986	10004	2986	10004	1500	180	18	4618	462	4618	455	78

FIG. 2—Example of dialog from application program for multiprobed surface-defected specimen with test machine control.

```
run
REVERSING DC POTENTIAL SYSTEM/4 HIGH-072882-1500

NUMBER OF SAMPLES [1=<N<=4]? 1
PRESTART/RESTART TIME [HR]? 0
PRINT/AVG TIME INTERVAL [HR]? .1

[CTL/R STARTS THE PROGRAM/CLOCK RUNNING]
[CTL/C STOPS PROGRAM EXECUTION]
```

TIME HRS E+0	NO DATA PTS E+0	AMPS (CH8) CT-1 E-3	SAMPLE-1* * * * * * * * * * * * * ACT(CHO)* * AVG NORM E-8 E-4		REF(CH1)* * AVG NORM E-8 E-4		NET RATIO E-4	COD-1 (CH9) VOLTS E-3
.1	298	3843	4277	10000	2822	10000	10000	51
.2	298	3842	4275	9995	2821	9999	9997	52
.3	298	3841	4274	9995	2820	9996	9999	52
.4	298	3841	4273	9992	2820	9994	9999	52
.5	298	3841	4273	9992	2819	9992	10000	52
.6	299	3841	4273	9990	2820	9994	9997	52
.7	299	3841	4273	9990	2819	9992	9998	52
.8	299	3840	4272	9989	2820	9993	9996	52

FIG. 3—*Example of dialog from application program for single-probed compact-type specimen for static testing.*

is valuable protection from a test running unmonitored as the result of the microcomputer "hanging" due to a power drop at the test facility.

The hardware comprising this technique is portable and rack mountable. Signal and control cables are BNC connections for ease of installation.

Specimen Design

Three types of test specimens have been used to date, CT, edge-notched rectangular bar, and surface-defected rectangular bar geometries. Some testing has been accomplished on internally defected short pipe sections with equally good results. All specimen geometrics share several common characteristics in applying the reversing d-c method for crack monitoring. Platinum wire has been selected for both current and potential leads because of its inert properties, ease of spot welding, and resistance to oxide formation in high-temperature aqueous environments. Current leads 0.762 mm (0.030 in.) in diameter and 0.254-mm-diameter (0.01-in.) potential leads are spot-welded to the specimen at predetermined locations prior to installing the specimen in the test facility. The current leads are kept short, approximately 203 mm (8 in.) because of the high resistance of platinum. The specimen wires are attached to permanently installed platinum "feed-through" entry leads of heavier-gage wire, typically 1.016 mm (0.040 in.) diameter for current and 0.508 mm (0.020 in.) diameter for potential. Outside the autoclave, the entry leads, which are kept to lengths of approximately 305 mm (12 in.) are attached to copper instrument leads. Tetrofluoroethylene sleeving, capable of withstanding 288°C (550°F) water without degradation, provides insulation for all platinum leads.

The CT specimen employs three pairs of leads: current, reference, and one active probe pair. The choice of positions for current and active probe potential leads is dictated by the need to establish a balance between maximum sensitivity of potential to crack growth and minimum sensitivity to any inaccuracy in placement of probes. The positions selected (Fig. 4) were based on the results of Klintworth [8], who conducted a finite-element analysis relating potential output at any point on the surface of a CT specimen to crack length for several current lead attachment locations. The reference probe position was selected to show the minimum change with crack growth.

Surface-defected specimens incorporate one pair of current leads and one pair of reference potential probes and from one to six pairs of active probe pairs. Figure 5 indicates relative positioning of the initial defect and probe pairs. Current leads are attached in the grip head area, assuring uniform current distribution in the gage section. The reference probe is well removed from the defect and specimen fillet and reacts only to temperature, current, and strain in the specimen gage section. Cracks whose shapes do not change during growth can be adequately characterized by a single active probe pair. When a single active probe pair is employed, it straddles the defect and is centered on its surface length. The probe spacing is again a compromise of sensitivity and accuracy of locating probe positions. Multiple active probe pairs are used to monitor crack growth when it cannot be assumed that the aspect ratio of the crack will remain constant. Use of several probe pairs permits measurement of size and shape of a growing crack. The most common configuration of probes employed is shown in Fig. 5. Other locations, including positions on the back side of the specimen, can also be used for making measurements.

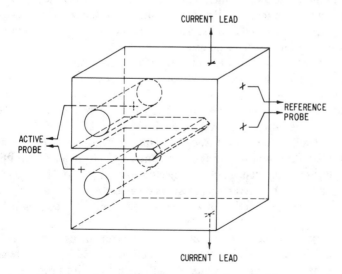

FIG. 4—*Probe configuration used in making reversing d-c potential measurement with compact-type specimen.*

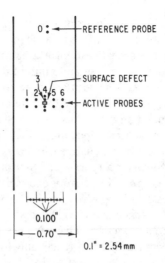

FIG. 5—*Probe configuration used in making reversing d-c potential measurement with surface-defected specimen.*

Test Procedure

Effective operation of a test facility incorporating the reversing d-c electrical potential method requires operating knowledge of the test equipment and familiarity with the dialog and diagnostics of the application programs. Application programs for all test specimen configurations require proper identification of probe pairs to amplifier channels. Current levels must be selected to produce microvolt levels which will not increase to instrument amplifier saturation during the test run. Prior to installing a test specimen, spacing of all probe pairs is measured and recorded. This spacing is used in analytical programs during and at the conclusion of testing.

While numerous programs are available for control of different types of tests and specimen geometries, examination of the dialog (Fig. 2) for a "K-controlled" test of a multiprobed, surface-defected specimen will illustrate the procedure to be followed in testing. The first line identifies the application program, then headings of MACHINE, SAMPLE, TEST, BLK SZ, and PROGRAM lead the operator through the dialog. Under MACHINE heading, LOAD RANGE inquires number of pounds equal to 10 V. TARE LOAD inquires what load the specimen is subjected to at the start of the test and allows for loads due to differential pressures created by vacuum or pressurization of environmental chamber or any desired load on the specimen prior to test initiation.

Because this example is a real-time, load-controlled test, the SAMPLE heading inquiries are needed to satisfy the algorithm used to calculate crack depth from the net ratio of the symmetrically located active probe pair. The thickness, original crack depth, aspect ratio, and probe spacing calibrate the start-up conditions. The TEST heading prompt determines an initial start-up or a sequence restart. A zero response to PRESTART/RESTART cues the program to store the first averaged microvolt value of each potential channel for normalization, whereas other than zero response cues the INITIAL OR REINITIALIZE prompt. If given a zero response for each channel, the first averaged microvolt values will be stored for future normalizing and the test has been reinitialized. If the original microvolt value of each channel is entered, original ratioing will be maintained. CYCLES FOR INITIALIZATION is negated when a zero response is entered. When the crack depth is questionable, a number of cycles of initialization should be entered. A prompt of STRESS GAIN AT START-UP will inquire at what percent of desired stress intensity the initializing cycles will be run. This action will protect overloading the specimen should the calculated crack length be greater than the PRESENT CRACK DEPTH entered. The last three inquiries under the heading TEST allow the operator to designate at what percentage of load amplitude should potential values be sampled. A zero response to DATA THRESHOLD allows sampling during the entire load amplitude whereas a 0.5 response allows sampling during the top 50% of load amplitude.

Empirically, noise levels have been reduced to an acceptable level when 50 or more groups of 16 readings are averaged. Figure 6, which displays data obtained on cyclically loading a surface-defected specimen of SA 333, illustrates the reduction in scatter obtained by such averaging. A least-squares best-fit line has been drawn through the blocked data points. Analysis of these data shows that the scatter is random and hence averaging is justified. The number of cycles needed to accomplish 800 (50 $\times$ 16) averaged readings to represent one data point is a function of waveform, period, and data threshold. The program calculates the minimum number of cycles needed to equal or exceed 800 readings and cues the operator. The operator is not required to use the cued block size; however, it is a guide to the level of averaging required to minimize noise in recorded data. Block size and number of blocks per printout do not have to be equal, but generally are. The final heading, PROGRAM, allows the operator to set up the test machine control based on constant load, constant K, programmed load, or programmed K, the latter two based on crack depth or number of cycles. A YES response brings up the prompt CONSTANT [L]OAD OR [K]. Either response will bring up three additional prompts: K{MAX}, K{MIN}, and A VALUE AT TERMINATION where A is crack depth. In constant K-mode, the load will decrease as the crack grows, whereas the initial loads will remain constant with crack growth in constant load mode. Both will terminate at the entered B-value. A NO response brings up an explanation of how to build a K or L table consisting of crack length, K_{max} and K_{min} entries per line of table. Twenty-five table entries are allowed and table entry is terminated when a zero is entered for K{MAX}. For operator convenience, the compiled table can be printed out

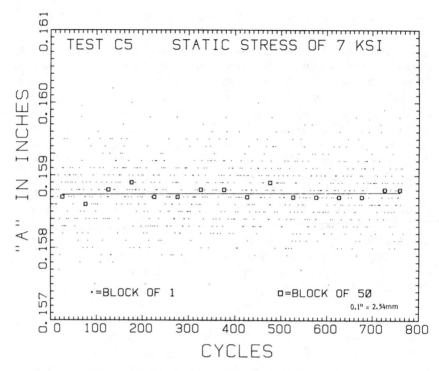

FIG. 6—*Minimization of scatter in data by block averaging.*

for examination prior to starting the test. Control/R starts the function generator and program execution. Control/H stops both whereas a Control/C stops program execution only. All cyclic data are printed out regardless of the specified block size unless the terminal's space bar is keyed, which limits the printout to blocked data only. Control/R will return the printout to all cyclic data. Typically, blocked data only are recorded and block size is adjusted so 0.0254 mm (0.001 in.) or less of crack growth occurs between blocks. The operator can change the block size, waveform, or period with a Control/I. Program execution stops and the prompts WAVE PERIOD through BLOCK SIZE dialog allow the appropriate changes. Control/R will restart the test at the new conditions.

Data recorded on magnetic tape can be removed from the terminal at any point in time for analysis, and hard copy inspection can be made for immediate status of the test.

Applications of Reversing D-C Potential Measuring

The reversing of d-c potential technique has been employed successfully in our laboratory for several years [2–7]. Several test results will be described to illustrate its applicability to testing under a wide variety of conditions, including static and cyclic crack growth in high temperature pressurized water, and use of CT and surface-defected specimens as well as structural shapes. The tests described have been selected because they illustrate some of the outstanding features of the test technique, such as use of a reference probe to provide temperature compensation, use of multiple probes to determine the shape of a crack front in surface-defected specimens, and the ability of multiple readings taken within a single cycle to differentiate between cycle-dependent and time-dependent crack growth and to show closure effects. Near-future testing includes crack measurements in service components in the laboratory and eventually in the field.

Temperature Compensation

While only a limited amount of testing was done on pipes, one such test demonstrates the capability of a reference probe to minimize the effect of temperature variations during testing [6]. A section of 10.16-cm-diameter (4 in.) SA 333 carbon steel pipe (nominal composition carbon 0.22, manganese 0.80, phosphorus 0.04, sulfur 0.05 maximum, balance iron) with a 5.080-mm-radius (0.200-in.) thumbnail defect at its inner wall was monitored for several days while subjected to pressurized, elevated-temperature water. No axial load was applied to the pipe section and no crack growth took place. Minor temperature fluctuations occurred early in the test but after 42 h the test was terminated and the heater turned off. Water continued to flow

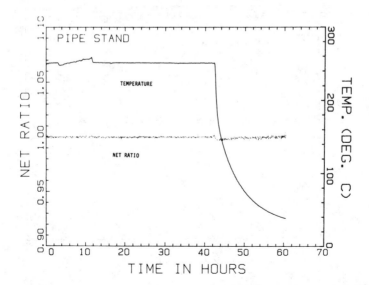

FIG. 7—*Effectiveness of reference probe in minimizing the influence of temperature variation during testing.*

through the pipe at a rate which caused a severe drop in temperature during the first two hours of cooling. Even during this period of cooling, the net ratio from the probe pair straddling the center of the effect remained almost constant (Fig. 7). The slight irregularity in the net ratio caused by this sudden change in temperature represents an indicated change in crack depth of approximately 0.00508 mm (0.0002 in.). After a period of a few hours when all temperature gradients had been eliminated, the net ratio returned to its original value.

Stability

A requirement of high sensitivity measurement is good long-term stability. Figure 8 is a plot of data from a CT specimen test run at 288°C (550°F) in pressurized water at a low cyclic stress level at 0.021 Hz. The probe configuration was as shown in Fig. 4. The least-squares best-fit line has zero slope where ±2% in net ratio represents approximately ±1.882-mm (0.0741-in.) change in crack depth.

Measurement Sensitivity

Figure 9 is a plot of data taken during testing of a surface-defected specimen of carbon steel in 149°C (300°F) oxygenated water using a probe configuration as shown in Fig. 5 [7]. The specimen was 17.8 mm (0.700 in.) wide

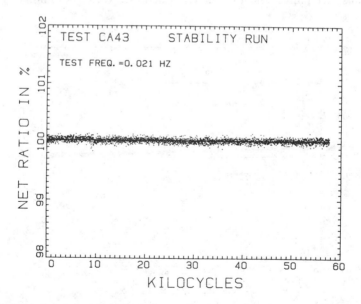

FIG. 8—*Stability and scatter in normalized potential values (net ratio) for 1T compact-type specimen at 4 A.*

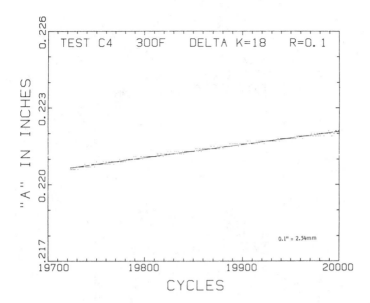

FIG. 9—*Sensitization of 0.00254 mm (0.0001 in.) in surface-defected specimen with cross-sectional area of 158.064 mm² (0.245 in.²) at 3 A.*

and 8.89 mm (0.350 in.) thick in the gage section and contained an initial defect of 3.81 mm (0.150 in.) radius at its midlength. The portion of the test depicted includes 300 cycles during which the crack increased in depth from approximately 5.601 to 5.639 mm (0.2205 to 0.2220 in.), a crack growth rate of 12.7×10^{-5} mm/cycle (5×10^{-6} in./cycle). From these data, it is clear that accurate crack growth rates can be determined with small amounts of crack growth, hence a single specimen can be employed to gain crack growth rates under a number of different conditions.

An additional benefit of high resolution of crack length is the ability to measure crack growth rates which are similar to those acceptable in service, hence decreasing the necessity of running accelerated tests and extrapolating data. Figure 10 shows measured crack growth rates approaching 2.54×10^{-6} mm/h (10^{-7} in./h) in a compact tension specimen of SA 533 pressure vessel steel subjected to static load conditions in pressurized elevated-temperature water of two levels of dissolved oxygen. A period of less than one month's testing time was required to obtain these data points

Measurements of Crack Shape

Multiple-probed surface-defected specimens have been used to demonstrate the capability of determining changes in crack depth and shape as crack growth proceeds [6–7]. The procedures used in testing an SA 333 car-

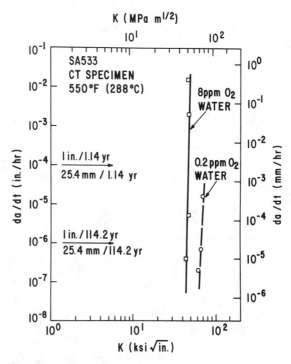

FIG. 10—*Comparison of crack growth rates as measured by reversing d-c potential technique in water with two different oxygen contents.*

bon steel specimen are shown with reference to Fig. 11. A crack with an initial aspect ratio of 0.25 and depth of 0.635 mm (0.025 in.) was probed with six probe pairs as shown in the upper left of the figure. The specimen was cyclically loaded in 288°C (550°F) oxygenated water to cause the crack to grow. The microvolt response of the several probes is plotted as a function of cycles in the upper right of Fig. 11. Note that the several curves are as would be expected. Probe pairs 3 and 4 show an immediate response as the crack grows, 3 showing a greater change than 4 since it is closer to the crack and is therefore more sensitive to crack growth. Probe pairs 2 and 5 show little response at first, but as the crack continues to grow, they begin to respond in a similar fashion since they are equidistant from the center of the crack. Probe pairs 1 and 6 are more remote and respond only slightly and only when the crack is very large. Probe pair 0, which is the reference probe, does not react to changes in defect size. Data obtained in this manner are used to calculate the dimensions of that semi-ellipse which most closely approximates the crack shape. A program has been devised which prints out the depth, length, and area of the crack at the end of each cycle or at the end of any desired number of cycles. In addition to these dimensions, it calculates the offset of the axis of

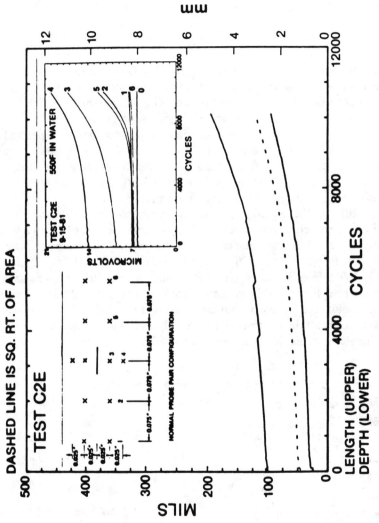

FIG. 11—Calculated crack depth, surface length, and area change in multiprobed surface-defected specimen.

the best fitting ellipse from the axis of the original defect, hence showing clearly any assymetry in growth. Figure 11 shows a plot of the length, depth, and square root of the area calculated from the voltages measured by the six active probe pairs. Examination of these dimensions shows that the crack has changed shape during growth. Its initial aspect ratio, that is, surface length/depth, of 0.25 changes gradually to 0.49 at the end of the test.

Cycle-by-Cycle Analysis

Data from a surface-defected specimen of SA 333 carbon steel subjected to cyclic loading in pressurized elevated-temperature water is shown in Fig. 12. This plot demonstrates the capability of making measurements which distinguish between time-dependent and cycle-dependent crack growth in a test using a trapezoidal waveform with one-hour maximum load hold period [2]. A constant slope line has been drawn through each of the maximum load hold data of seven cycles. The offset from one line segment to the next is a result of cyclic crack growth. The time-dependent growth is greater than cyclic growth for this material subjected to this testing condition.

The ability of this technique to make many measurements during a single low-frequency cycle permits making observations which may lead to a better understanding of the mechanism of crack growth in a wide range of environments [4]. Results obtained from testing a CT specimen of SA 333 carbon steel in 288°C (550°F) pressurized water under two conditions can be used to illustrate this. Figure 13 shows the voltage response obtained during five 80-min sinusoidal cycles with R (min load/max load) = 0.1, while Fig. 14 shows similar results attained during five cycles when R = 0.5. Examination of the data for either of the five cycle periods shows that the response to loading and

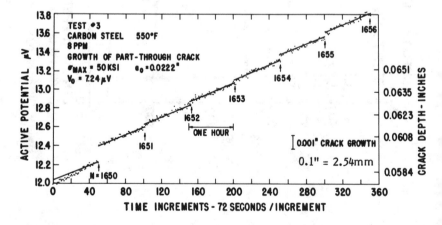

FIG. 12—*Cycle-by-cycle analysis distinguishing between time-dependent and cycle-dependent crack growth in a surface-defected specimen using trapezoidal waveform.*

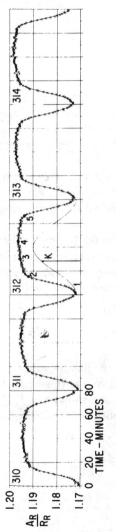

FIG. 13—*Electrical potential response—active/reference ratio—for five successive cycles with smoothing curve-sinusoidal loading, R = 0.1.*

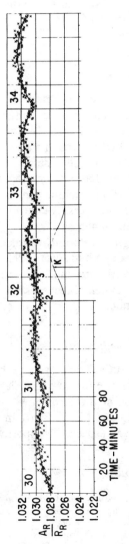

FIG. 14—*Electrical potential response—active ratio/reference ratio—for five successive cycles with smoothing curve-sinusoidal loading, R = 0.5.*

unloading is identical in each cycle. The only difference is an upward offset indicating that growth is taking place. The dependence of potential on load suggests that closure takes place as the load is decreased. The amount of closure is seen to be much larger when R = 0.1 than when R = 0.5. It will further be noted in each figure that the maximum voltage is attained in each cycle at a point in time after the load has started to drop, suggesting that growth continues as long as the load is maintained at a level fairly close to the maximum.

Conclusion

The reversing d-c electrical potential method is a valuable tool for making measurements of crack initiation and crack growth rates under a wide range of testing conditions. Its applicability with CT, surface-defected, and edge-notched specimens in static and cyclic testing under difficult environmental conditions has been demonstrated.

Software has been developed for data acquisition with static and cyclic testing. Testing machine control is available for cyclic testing with a variety of input parameters. All hardware has been designed to adapt to a wide variety of testing machines.

Long-term stability coupled with a ±0.0025-mm (±0.0001-in.) sensitivity in crack depth in surface-defected specimens has permitted measurements of low growth rates in short testing time, thus allowing many test conditions per test specimen. Mechanistics studies of crack growth are feasible with this technique.

Acknowledgments

The authors acknowledge the support and encouragement of both the General Electric Co. and the Electric Power Research Institute in this work. In the case of the former, we particularly acknowledge the support of R. F. Berning and M. G. Benz. In the latter case, R. L. Jones and J. D. Gilman have been especially supportive of this activity.

References

[1] *The Measurement of Crack Length and Shape During Fracture and Fatigue*, C. J. Beevers, Ed., Engineering Materials Advisory Services, Ltd., The Chameleon Press, London, 1980.

[2] Catlin, W. R., Morgan, H. M., Lord, D. C., and Coffin, L. F., "A Method for Monitoring Crack Growth from Flaws in Piping and Other Components," *Proceedings*: Seminar on Countermeasures for Pipe Cracking in BWR's, Electric Power Research Institute Report EPRI WS-79-174, Palo Alto, CA, Vol. 2, May 1980.

[3] Prater, T. A. and Coffin, L. F., "Part-Through and Compact Tension Corrosion Fatigue Crack Growth Behavior of Carbon Steel in High-Temperature Water," *Environmental Degradation of Engineering Materials in Aggressive Environments, Proceedings*, Second International Conference on Environmental Degradation of Engineering Materials, Virginia Polytechnic Institute, Blacksburg, VA, Sept. 1981, pp. 59–72.

[4] Prater, T. A. and Coffin, L. F., "Experimental Cycle Analysis of Fatigue Crack Growth in Compact Type Specimen Geometries," General Electric Co. Report No. 82CRD039, Schenectady, NY, Feb. 1982.

[5] Prater, T. A., Catlin, W. R., and Coffin, L. F., "Surface Crack Growth in High Temperature Water," *Proceedings*, International Symposium on Environmental Degradation of Materials in Nuclear Power Systems—Water Reactors, Aug. 22–24, Myrtle Beach, SC, Aug. 22–24, 1983, to be published.

[6] Prater, T. A., Catlin, W. R., and Coffin, L. F., "Environmental Crack Growth Measurement Techniques," Electric Power Research Institute Report EPRI NP-2641, Project 2006-3, General Electric Co. Research and Development, Schenectady, NY, Nov. 1982.

[7] Prater, T. A., Catlin, W. R., and Coffin, L. F., "Environmental Crack Growth Measurement Techniques," Electric Power Research Institute Project 2006-3, General Electric Corporate Research and Development, to be published.

[8] Klintworth, G. C., "Fatigue Crack Propagation in High Strength Low Alloy Steel Using an Electrical Potential Method," MS thesis, Imperial College of Science and Technology, University of London, 1977.

David A. Topp[1] and W. D. Dover[1]

Crack Shape Monitoring Using A-C Field Measurements

REFERENCE: Topp, D. A. and Dover, W. D., "Crack Shape Monitoring Using A-C Field Measurements," *Automated Test Methods for Fracture and Fatigue Crack Growth, ASTM STP 877*, W. H. Cullen, R. W. Landgraf, L. R. Kaisand, and J. H. Underwood, Eds., American Society for Testing and Materials, Philadelphia, 1985, pp. 86–100.

ABSTRACT: A multi-central processing unit (CPU) computer network has been installed at University College London for fatigue tests and automatic inspection work. This system is described, together with a newly developed alternating-current field measurement inspection system for crack measurement, in use on some novel stress corrosion tests. The complete system makes it possible to conduct tests where the crack shape evolution can be monitored.

KEY WORDS: crack measurement, crack shape, fatigue crack growth, fatigue testing

The London Centre for Marine Technology (LCMT) at University College London (UCL) has for some years been actively involved in the large-scale fatigue testing of tubular welded joints as used in the offshore industry. Other recent projects include a fatigue and fracture mechanics study of large threaded tether connections for use in the new generation of tension leg platforms, the inspection of turbine disks, and fatigue and fracture testing of high-strength bolts used in the space shuttle.

Much of this work involves fracture mechanics studies of the fatigue crack growth behavior. In order to fully assess this behavior it is necessary to determine not only the crack depth but also the shape of the crack. Crack shape evolution gives valuable insight into the mechanisms by which the cracks progress in these often complex geometries, under cyclic and random loading.

The fatigue test facilities at LCMT have capacities ranging from 100 to 2500 kN, the majority of these being servo-hydraulic machines including

---

[1]Research assistant and professor of mechanical engineering, respectively, London Centre for Marine Technology, University College London, Torrington Place, London, England.

standard test frames and purpose-built rigs for the testing of welded tubular joints. The need for automated and semi-automated testing was recognized some years ago and, to facilitate this, a number of test rigs were interfaced with PDP II minicomputers that were used mostly for real-time load generation and data collection, including crack depth measurement using an alternating-current (a-c) potential-drop method developed in-house but now available commercially as the Crack Microgauge.[2] The pressure for further automation has now led to the introduction of a new real-time multiuser computer facility, for automated fatigue testing and inspection, with availability on a number of sites. These innovations are described together with a recent example of stress corrosion tests using these techniques.

Crack Measurement Using A-C Potential-Drop Techniques

The measurement of cracks using a-c potential-drop techniques (acpd) has been described previously [1–3]. In this work it was shown that both detection and monitoring were possible in both air and seawater. Originally the technique involved the use of a hand-held probe that was manually traversed across the area of interest. Crack depths could then be calculated from two readings, one on an uncracked region close to the crack and the second taken as the probe spans the crack. In practice, however, if a large number of points are to be inspected this process can be time-consuming. In addition, a hand-held probe is difficult to reposition in exactly the same place for subsequent measurements; this can be important if the crack shape evolution is to be accurately determined. Another obvious problem is that for some geometries it is not physically possible to access the crack using a hand-held probe.

To overcome these problems, two types of automated systems have been developed to enable the crack shape to be measured.

Measurement of Cracks Using Fixed Probes

In some test situations the location of the fatigue crack is known; for example, the growth of fatigue cracks from notches or stress concentrations. Fatigue crack growth in tubular welded joints is a case where the crack initiation site could be predicted. For these cases fixed probes can be spot-welded to the specimen in the expected crack initiation zone. The probes are typically 6 mm long and 0.5 mm in diameter and are positioned in pairs. The pairs of contacts (effectively forming probes) are hard-wired to the a-c instrument via a special-purpose switch unit. This switch unit is specifically designed to deal with the very small voltage measurements and to be under computer control. A range of probes up to 64 in number has been used for tubular joints, and when scanned sequentially they determine the crack depth at adjacent sites

---

[2]Crack Microgauge is produced by the Unit Inspection Co., Swansea, Wales.

and effectively give a scan along the crack. From these readings the crack shape is derived and can be stored in the computer for comparison with subsequent scans. Figure 1 shows an example of recorded data taken during a test on a tubular welded Y-joint. These results show how a crack initiates at a number of sites and subsequently coalesces to form a single large crack. Although this process gives only one final crack, the individual crack fronts can still be discerned well after formation of the single dominant crack.

Using this technique, both specimen loading and inspection can be under computer control. Thus any test load sequence can be applied and the corresponding crack shape evolution recorded. The technique can also be used to obtain material data using precracked specimens. Fixed probes can be spot-welded into the starter notch of the specimen and the crack growth monitored at a number of points along the crack front. Using this method constant K-tests can be carried out with the computer adjusting the specimen loading to take into account the crack depth at any time.

Crack Measurement Using Movable Probe

For some situations it is necessary to be able to locate, as well as characterize, the crack. To accomplish this, a technique has been developed that involves moving a probe across the specimen surface in a controlled way. Once

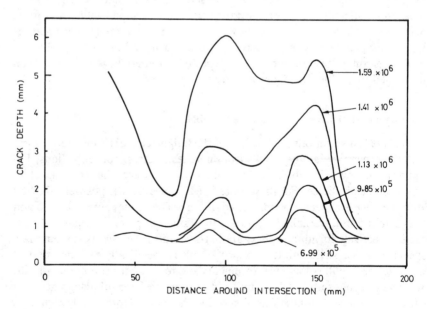

FIG. 1—*Weld toe fatigue crack profiles recorded at various stages during a random load fatigue test on a welded tubular Y-joint.*

the crack has been detected it can then be characterized by determining its depth at various locations along its length.

For flat plates the plate remains fixed with the a-c field connected to it. One system used is to track the probe across the plate using stepping motors under computer control.

For threads or turbine disks the probe remains fixed in two planes and is traversed along the specimen as the specimen rotates. This time the specimen rotation is under computer control, again using a stepping motor drive. The probe path is a helix, which is ideal for thread inspection, each revolution of the specimen advancing the probe by one tooth pitch. In the case of a large thread, the distance inspected by the probe can be as much as 8 m. This is initially inspected rapidly. If a crack is detected the probe is then returned for a more detailed inspection of the area. Typical plots of crack shape are shown in Fig. 2. Again these results show the way in which the crack has initiated at a number of sites and subsequently merged to form one large crack.

In the case of threads and turbine disks, one problem experienced during inspection is variable contact at the field inputs as the specimen rotates. Also, uniformity of the input field can be a problem in these geometries. This has been overcome by the use of an induced field instead of the conventional im-

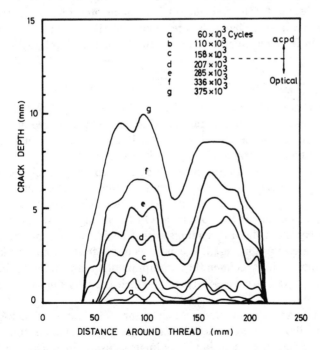

FIG. 2—*Crack shape evolution for a fatigue crack located at the thread root of a casing joint (offshore riser-type connection).*

pressed field. A coil is attached to the probe and the a-c field is passed through this coil. This induces a field locally into the specimen at the point of interest, which is then measured using the probe. Thus the field at the region of interest is always similar regardless of the surrounding geometry.

Interpretation of A-C Readings in Terms of Crack Depth

One of the advantages of this new method for a-c field measurements is that the true surface voltage is recorded and this can be interpreted in terms of crack depth by using theoretical solutions of the field distributions. In many situations the field can be assumed to be linear, in which case the measured voltages v_1 and v_2 taken just off and just over the crack can be interpreted through the following equation in terms of crack depth (d_1)

$$d_1 = \left(\frac{v_2}{v_1} - 1\right)\frac{\Delta}{2} \tag{1}$$

where Δ is the probe contact spacing.

For nonuniform fields the irregular distribution needs to be calculated, and for several cases this has already been done [4]. Using the theoretical nonuniform field solutions [4] the crack depth can still be interpreted in terms of the two measured voltages. The procedure adopted has been to retain the calculation shown in Eq 1 but in addition to calculate the true depth (d_2) using

$$d_2 = Md_1 \tag{2}$$

where M is a modification factor to account for field nonuniformity and is available from a computer program or table for the specific geometry under study. The two approaches can be illustrated by some results taken in tests on two different types of fatigue crack growth as described below.

The uniform field interpretation was used for a compact tension specimen (50 mm thick) subjected to constant-amplitude cycling in a seawater environment. The measurements were taken over a crack increment of 25 mm and were on the centerline of the specimen. On fracturing the specimen it was possible to correlate the a-c readings with fracture surface markings and these results are shown in Fig. 3. It can be seen that for the total crack increment of 25 mm in this specimen the assumption of a uniform field is justified and the scatter was at most 3% (part of which is surely due to the optical measurements).

The second example involved measurement of the fatigue crack growth from a semi-elliptical notch. Here it was necessary to calculate first d_1 and then using the theoretical solution [4] estimate d_2 using Eq 2. The results, for both d_1 and d_2, from several tests were again compared with fracture surface markings and are displayed in Fig. 4. It can be seen that with the aid of the

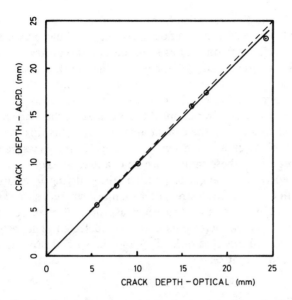

FIG. 3—*Predicted and optically measured (after fracture) crack lengths for a 50-mm-thick compact tension specimen (using Eq 1).*

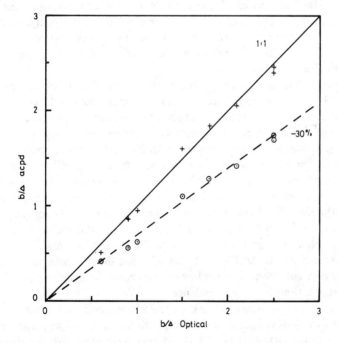

FIG. 4—*Predicted and optically measured (after fracture) crack depths for a semi-elliptical surface crack in a plate specimen (using Eq 2) (unmodified results using Eq 1 also shown).*

theoretical solution the crack depth can be accurately estimated even for the semi-elliptical case. So far, in all of our fatigue tests, it has proved possible to evaluate the crack depth and it has been found unnecessary to conduct any calibration. It should be recognized, however, that for these tests the nominal stresses are often low and varying between fixed limits. For this reason it has been found unnecessary to consider any effects due to piezomagnetism. For many steels the presence of mechanical stress can cause the magnetic permeability to be changed [5]. In the case of a-c measurements this would mean a change in "skin" thickness and hence a change in the measured potential drop. For most metals these changes are small and can be considered as negligible compared to the changes in path length occurring in a fatigue crack growth test. In contrast, for a stress corrosion test where a slow ramp load is used, the change in stress can produce a measurable effect especially in situations where small crack increments need to be detected. The tests to be described later involve this type of loading and the special procedure adopted will be described in that section.

Computer System

Initially the Fatigue Laboratory at UCL was equipped with several DEC PDP 11 series minicomputers suitable for real-time control of laboratory experiments. Hard disk units with removable packs were fitted to all the computers and the main use was for signal generation and data logging. Since this time the need has arisen for an increase in the number of sites to be serviced with computer facilities. In addition, the costs of maintaining the original system has risen steadily.

The solution adopted for the problem was to build a multi-central processing unit (CPU) network [6] where only one computer (the host) has mass storage peripherals as a necessity. All other CPUs on the system (satellites) are connected to the host through high-speed interfaces operating in the megaband range through multicore cables up to 200 m in length. Satellite CPUs are fitted with only those peripherals necessary for their particular task. This arrangement was possible through the use of a software package "Star Eleven" developed and marketed by Hammond Software. This package is based on the DEC RT-11 operating system which we were already using. The advantages seen for this system lay in the switch from the use of several mass storage peripherals to just one. These items have been the main problem in cost and reliability. In addition, it was felt that the system opened up the possibility of a common software library that could be accessed by all users while retaining their own independence.

So far seven satellites have been connected to the host and these are being used for a variety of fatigue and stress corrosion test as well as inspection for cracks. A software package (FLAPS) has been written for use at these satellites and this consists of three main programs, design or preprocess, run, and post-process. These permit the user to interactively select his experimental setup using a menu of options of the type shown below:

Design Program

TO ENTER INITIAL PARAMETERS	TYPE 1
TO SELECT SIGNAL GENERATION	TYPE 2
TO SELECT REAL TIME CLOCK	TYPE 3
TO SELECT SAMPLING OPTIONS	TYPE 4
TO SAVE SELECTIONS ON FILE	TYPE 5
TO READ OLD CONTROL FILE	TYPE 6
TO RUN CONTROL PROGRAM	TYPE 7
TO RUN GRAPHICS PROGRAM	TYPE 8
TO REPRINT OPTION TABLE	TYPE 9
OR TO EXIT FMT1 (FINISHED)	TYPE 10

Selecting Option 2 would give the following new menu:

Signal

IF EXTERNAL SIGNAL REQUIRED	TYPE 1
TO SELECT SQUAREWAVE SIGNAL	TYPE 2
TO SELECT RAMP SIGNAL	TYPE 3
TO SELECT SINE WAVE SIGNAL	TYPE 4
TO SELECT RANDOM SIGNAL	TYPE 5
TO SELECT BLOCK LOADING	TYPE 6
TO SELECT INSPECTION SIGNAL	TYPE 7

Option 7 here contains a further set of options all related to crack measurement with the following menu:

Inspection

TO SELECT CTS INSPECTION	TYPE 1
TO SELECT SEN INSPECTION	TYPE 2
TO SELECT TETHER INSPECTION	TYPE 3
TO SELECT DISK INSPECTION	TYPE 4
TO SELECT TUBULAR JOINT INSPECTION	TYPE 5
TO SELECT T-BUTT WELD INSPECTION	TYPE 6

Selecting Option 2 for example would give the program required for the stress corrosion test to be described later.

Nature of Stress Corrosion Test

Stress corrosion cracking data were required on a high-strength steel of section size 850 by 160 by 75 mm. The expected environmental loading indicated that a slow ramp load test with a superimposed sine wave loading at 1/6 Hz and amplitude of 1% of the ramp load would be most appropriate. Stress corrosion threshold and crack growth data were required.

Two main requirements governed the choice of the experimental setup. These were the need for a very small dynamic amplitude, to be applied for a long period, and the need for a very sensitive crack measurement and monitoring system. In addition, the maximum force required during the test was 800 kN. These requirements were met as described in the following.

Loading System

The first system considered was a servo-hydraulic test machine of 1000-kN capacity. However, a stiff specimen with very slow ramp loading means the servo valve is hardly moving, which can lead to buildup of sediment on and around the valve spool. This is known as silting and results in erratic valve movements since the valve effectively sticks until there is a sufficient demand to overcome the sticking, upon which the valve responds instantaneously to the new demand level. The temporary loss of control of the valve produces a step loading with overshoots, as opposed to a slow ramp, which was unacceptable. A further disadvantage is that servo-hydraulic machines have a fast response. In the present work this was not required and, although use of a hydraulic slice could reduce this response, it was felt that given also the cost of running power packs continuously for several weeks, an alternative solution should be sought.

The alternative considered was a servo-electric machine. Unfortunately these are not available with capacities up to 800 kN, but the advantages of low-power consumption and the ability to maintain the load on the specimen even after power failure indicated that this system merited further consideration. This led to the development of an 800-kN test rig built around a 100-kN Instron servo-electric machine. The details are given below. To overcome the problem of load capacity, a hydraulic load intensifier and an external test frame were designed. The test rig is shown in Figs. 5 and 6.

The intensifier comprised three steel flat jackets of the type commonly used for bridge installations. Two small jacks were inflated with hydraulic oil and placed in the test machine. These were then connected hydraulically to a larger jack in the test rig. The jack in the test rig loaded the specimen through a 1000-kN load cell which was connected to the machine controller. (The load cell in the test machine was disconnected.) Using the flat jacks provided a closed hydraulic system with no possibility of leaks other than at the hydraulic hose connections. During the test the effective area of the jack changes with

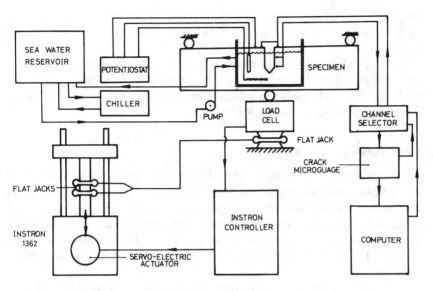

FIG. 5—*Schematic diagram of the stress corrosion test system based on the servo-electric actuator and flat jacks.*

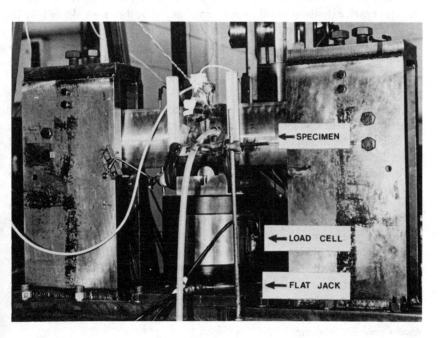

FIG. 6—*800-kN stress corrosion test rig (the 800-kN flat jack is located beneath the load cell).*

displacement and thus the load amplification changes. This is not important as the test load is governed by the load cell positioned next to the test specimen.

In this configuration the demand signal is fed into the system in the usual way and the electric actuator responds until the load at the specimen is achieved. This fully load-controlled system satisfied all the loading requirements and consumed only 2 kW of power.

Stress Corrosion Test Procedure

The specimens were to be machined from large forgings and were tested under three-point bending. It was desirable to test as large a section as possible; however, the fact that these were cut from an existing forging meant that the maximum dimensions were predetermined. These considerations resulted in the specimens not conforming to the ASTM Test Method for Plane-Strain Fracture Toughness of Metallic Materials (E 399-83) standard for three-point bend tests.

A precrack was required in the specimens and this was produced in air at as low a ΔK as possible. To enable this to be carried out a chevron starter notch was machined in the specimen. The starter notch plus subsequent fatigue precrack conformed in all respects to the recommendation in E 399. The precracking was done in a conventional servo-hydraulic machine. It was apparent that crack extension could be very small and that the crack shape may be irregular. For this reason the specification called for continuous crack monitoring at the sides and center of the specimen. Since we are looking at thresholds, it is critical that loads be maintained at a high degree of accuracy since any overloads will cause blunting of the crack and obviously invalidate the results.

Clearly there were two specific constraints on this test. The first was that small crack extensions should be reliably measured both on the surface and at the center of the specimen. Secondly, the loading should be closely controlled to ensure that the predetermined load history could be achieved.

For crack measurement a system using spot-welded probes was used to measure the crack extension at both the center and side. Optical readings were also taken on the side of the specimen to confirm the a-c readings. In order to achieve the best resolution in the measuring system it was necessary to fix the probes as near to the crack as possible. To facilitate this the chevron of the starter notch was machined out after precracking.

A computer-controlled switching system was used to monitor the crack at the center and edge. A further switching device was introduced to enable the a-c field to be switched to the specimen only while readings were being taken. This was in order to prevent the a-c field affecting the impressed current corrosion protection levels.

Results

The raw data from the tests were collected in the form of a-c field voltage measurements across pairs of probes. An estimate of crack depth can be calculated, knowing the probe spacing and field voltages, using the simple formula described earlier. It was decided that in these tests the crack depth would be calculated initially using Eq 1 but that an overall correction factor would be applied to the measured crack growth increment at the end of the test once the actual crack increment had been measured on the fracture surface. This would account for variation due to the specific test conditions. In order to ensure that this procedure was valid, results obtained from the a-c technique on the surface of the specimen were compared with optical readings taken during the test at the same point. As mentioned earlier, it was expected that the a-c field measurement would contain an effect due to the changes in stress. This, if present, causes the crack depth, calculated using the simple formula, to apparently decrease. The effect was eliminated using the following procedure:

1. Crack depths were calculated using Eq 1 to obtain crack growth increments with time.
2. Crack growth increments were adjusted for the stress effect as described below.
3. A correction factor was applied to the crack growth increments. This was an experimental calibration factor to correlate the total crack extension predicted by acpd with the actual crack growth measured on the fracture surface. Figure 7 shows the crack growth curve calculated from the raw data. The fact that the crack length appears to decrease is indicative of the stress effect described earlier. The dashed-line extrapolation is interpreted as the initial crack depth, and crack growth is measured with respect to this datum.

Figure 8 shows the final crack growth curve, this time in terms of the absolute crack depth, and the comparison with optical readings. Good agreement is seen between optical and a-c measurements, confirming that the technique is acceptable. The data reduction technique was adopted for all crack measurements, and the results presented here were typical for the test series.

Conclusions

Recent advances in automated crack measurement using a-c potential-drop techniques together with minicomputers enable fatigue data to be obtained by automated test methods. A nonstandard test series described here shows the way in which these automated testing techniques can be used in both standard and nonstandard test configurations to obtain reliable crack growth data.

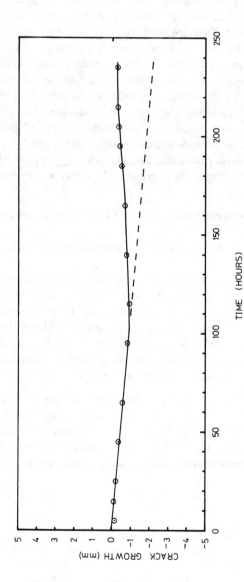

FIG. 7—*Five hourly averages of crack length measured on a three-point bend specimen during the stress corrosion test by means of acpd (the dashed line is the extrapolation of the effect on readings ascribed to the ramp load).*

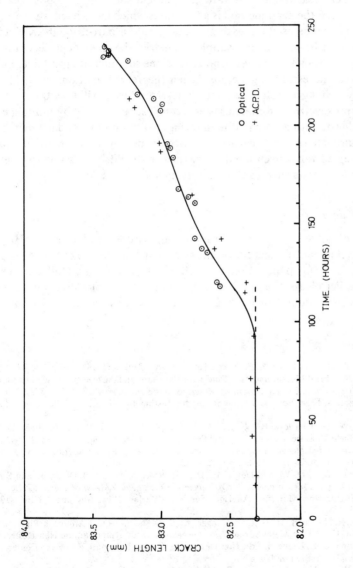

FIG. 8—*Calibrated acpd readings and optical readings for the stress corrosion test data shown in Fig. 7.*

The present standards and codes do not specifically include automated testing as an acceptable testing method. If standards are going to be updated to include automated test methods, the most important area will be that of crack measurement. The greatest change must therefore be the move from optical crack measurements toward automatic monitoring methods. Our experience is that the acpd method is a technique which lends itself well to automation and will be used increasingly as the need for part-through crack tests becomes more important. For simple geometries the crack depth is obtained using Eq 1. For more complex geometries some solutions are available, but it may become necessary to generate further theoretical solutions.

Clearly a changeover to automated crack monitoring will take time. Confidence in prospective automated systems can best be achieved by running conventional tests with automated monitoring in addition to standard methods. When automated crack measurements, confirmed by optical methods, have been accepted for conventional tests, then it should be possible to look at more complex tests such as those involving surface cracks.

Acknowledgments

The authors wish to thank the following venture partners in the Hutton Field Development Project for permission to include details of the stress corrosion tests in this paper: Conoco (U.K.) Ltd., Britoil plc, Gulf Oil Corp., Amerada Petroleum Corp. (U.K.), Amoco (U.K.) Exploration Co., Enterprise Oil Ltd., Mobil North Sea Ltd., and Texas Eastern North Sea Inc.

References

[1] Dover, W. D., Charlesworth, F. D. W., Taylor, K. A., Collins, R. and Michael, D. H., "The Use of A.C. Field Measurements to Determine the Shape and the Size of a Crack in a Metal" in *Eddy Current Characterization of Materials and Structures, ASTM STP 722*, G. Birnbaum and G. Free, Eds., American Society for Testing and Materials, Philadelphia, 1981, pp. 401–427.

[2] Dover, W. D., Charlesworth, F. D. W., Taylor, K. A., Collins, R. and Michael, D. H., "A.C. Field Measurement—Theory and Practice" in *The Measurement of Crack Length and Shape during Fatigue and Fracture*, C. J. Beevers, Ed., Engineering Advisory Service, Warley, West Midlands, U.K., 1980, pp. 220–260.

[3] Charlesworth, F. D. W. and Dover, W. D., "Some Aspects of Crack Detection and Sizing Using A.C. Field Measurements" in *Advances in Crack Length Measurement*, C. J. Beevers, Ed., Engineering Materials Advisory Services, Warley, West Midlands, U.K., 1982, pp. 253–276.

[4] Collins, R., Michael, D. H., and Ranger, K. B., "The a.c. Field Around a Plane Semi-Elliptical Crack in a Metal Surface" in *Proceedings*, 13th Symposium on Non-Destructive Evaluation, B. E. Leonard, Ed., Non-Destructive Testing Information Analysis Center, San Antonio, TX, 1981, pp. 470–479.

[5] Dover, W. D., Lugg, M., Collins, R., Michael, D. H., and Taylor, K. A., "Crack Monitoring Using a.c. Field Measurement," Contract Report for Ministry of Defence (N), University College London Internal Report No. MED/83/5, U.K., 1983.

[6] Broome, D., Dharmavasan, S., Dover, W. D., and Godfrey, L. J., "The Use of Digital Techniques in the Large-Scale Testing of Tubular Joints" in *Proceedings*, SEECO 83, International Conference on Digital Techniques in Fatigue, London, 1983.

Patrick M. Sooley[1] and David W. Hoeppner[1]

A Low-Cost Microprocessor-Based Data Acquisition and Control System for Fatigue Crack Growth Testing

REFERENCE: Sooley, P. M. and Hoeppner, D. W., "A Low-Cost Microprocessor-Based Data Acquisition and Control System for Fatigue Crack Growth Testing," *Automated Test Methods for Fracture and Fatigue Crack Growth, ASTM STP 877*, W. H. Cullen, R. W. Landgraf, L. R. Kaisand, and J. H. Underwood, Eds., American Society for Testing and Materials, Philadelphia, 1985, pp. 101–117.

ABSTRACT: The requirements for obtaining economic fatigue crack growth data from the threshold regime to instability in inert, gaseous, elevated-temperature, and aqueous environments necessitate the development of remote crack growth monitoring techniques. Two of these processes used extensively are the compliance method and the potential-drop method.

The performance of a microprocessor-based machine controller and data acquisition system utilizing the d-c potential-drop method of monitoring crack length is presented. The development of software for performing K-increasing and threshold tests and their compliance with the present ASTM standards are discussed. The successful development and use of the machine are shown by the results obtained during round-the-clock testing for acquiring crack growth data for a titanium alloy.

KEY WORDS: fatigue (materials), crack propagation, threshold stress-intensity factor, computers, test equipment, fatigue tests, titanium alloys

The increased requirements for the use of predictive methods to determine the fail-safe life of aircraft, gas turbine engines, nuclear power equipment, ground transportations vehicles, and ships has necessitated the development of more efficient generation of fatigue crack growth data. These data must be known for a wide range of materials such as aluminum alloys, low-alloy steels, nickel-based superalloys, and titanium alloys from the so-called threshold

[1]MTS graduate research fellow and Cockburn professor of engineering design, respectively, Structural Integrity Fatigue, and Fracture Research Laboratories (SIFFRL), Department of Mechanical Engineering, University of Toronto, Toronto, Ont., Canada.

stress-intensity range of fatigue crack growth, ΔK_{th}, to the instability point of fast fracture, ΔK_B. The characteristics and use of ΔK_{th} have been the subject of a recent review [1]. It has now been generally accepted as a design quantity although exact definition has still not been agreed upon. Bucci [2] discussed several possible methods of designating ΔK_{th}.

The manual control of a threshold test requires that an operator be present to monitor the load on the specimen and the crack length. One test under typical test conditions may take from two to five days of round-the-clock testing. The manpower requirements of supplying technicians for such purposes can prove to be exorbitant. Operator fatigue may also result in poor data acquisition, which may influence the overall reliability of the data obtained.

The need to perform fatigue crack growth tests under conditions where the crack cannot be monitored by visual methods led to the development of remote monitoring techniques. The potential-drop (PD) method utilizing either direct current (dc) or alternating current (ac) is one method presently in use. Originally the data from a system utilizing this method were obtained by such means as a strip chart recorder or by technicians monitoring voltage readout devices. The disadvantage of such methods were:

1. high manpower costs,
2. inefficiency, and
3. lack of up-to-the-minute data readout.

The problem regarding lack of instantaneous data readout is important. The information from a PD test is in the form of voltage signals which must be converted into a crack length measurement. Such a system still requires extensive technician time which must be added into the overall costs of running crack growth tests. It would be desirable to have a system that automatically calculates fatigue crack growth rate and the varying stress intensity, ΔK, or some other fracture mechanics parameter as the test is in progress. This would allow instant decision-making regarding whether or not a test should proceed or be curtailed.

Use of Computers in Fatigue

The development of the stand-alone mainframe computer in the late 1950s and 1960s and the minicomputer and microcomputer later on has greatly influenced the method of acquiring fatigue crack growth rate data. The coupling of a computer and a remote monitoring system allowed fatigue tests to be performed almost completely automatically without an operator being present.

One of the drawbacks to such a system utilizing a mainframe or minicomputer is cost. Extensive use of a DEC MINC 2 at the University of Toronto's Structural Integrity Fatigue, and Fracture Research Laboratories (SIFRRL) has revealed some shortcomings. The computer may only be utilized to control and acquire data from a single fatigue machine. Such a system used in a fatigue machine control and data acquisition mode is highly underutilized during the major portion of a typical test. Most crack growth tests require a sinusoidal waveform output for load, stroke, or strain control purposes. Typically, in a system where the computer uses one digital-to-analog (D-A) channel for the fatigue machine control signal, for a major portion of the time, the system is acting as a very expensive function generator. It is utilizing only a very small part of an expensive computer system.

The most recent developments in the microcomputer market have resulted in a number of inexpensive systems that could be expanded for use as a fatigue testing system at a fraction of the cost of a minicomputer system. The use of one of these in a PD system with dc as a combined machine controller and data acquisition system is outlined. Reasons for choosing the specific microcomputer are discussed as well as the various features built into the system.

The development of software for performing K-increasing and threshold tests and their compliance with present ASTM standards are also presented. The development and use of the machine is shown by results obtained during round-the-clock testing to acquire fatigue crack growth rate data on a titanium alloy.

Experimental Procedure

Equipment

The requirement to provide crack growth data on a number of titanium alloys for a gas turbine engine manufacturer necessitated that a system be acquired for control and data acquisition from a fatigue machine. It was decided that a microprocessor system be developed to allow for control of a closed-loop electrohydraulic servocontrolled fatigue test machine and for obtaining crack growth data by use of the PD method.

The system was designed around the Sinclair ZX-81 microcomputer, also known as the Timex-Sinclair 1000. This is the least expensive computer available today which utilizes BASIC as its operating language. The ability to market the computer for such a low price lies in its simple design. It utilizes a large-scale integrated circuit chip which replaces about 20 integrated circuits (ICs) from a previous design. The circuitry consists of only three ICs excluding the memory components.

The ZX-81 was expanded to a parallel bus based system to allow the installation of extra circuitry boards. A brief description of the circuit boards developed for the present system follows:

1. 16 K memory board: The ZX-81 includes only 1 K of memory. Previous experience indicated that a threshold program could be easily contained within 16 K of memory.

2. Four analog-to-digital input channels, 12-bit accuracy: The system required the ability to measure voltage level signals off the load cell, the PD across the specimen, and the d-c level. The fourth channel is not now used.

3. Two digital-to-analog channels, 12-bit accuracy: These were designed into the system for future expansion such as a control signal for a high-temperature oven. One channel is presently used for turning the dc to the specimen on and off.

4. Two-channel high-gain amplifier: To measure the low-level PD signal with the computer, it must be amplified before being input to the A-D board. The Channel 1 amplifier is presently set at 3500 for the PD signal. Channel 2 has a gain of 500 to monitor the current level flowing through the specimen.

5. Waveform generator board: This contains an adjustable ± 10 V sine wave oscillator which can be set to between 10 and 80 Hz and which is used as the fatigue machine control signal. Computer commands can adjust this signal to give different mean levels and span settings.

6. Cycle counter board: A counter was required in order to determine the number of fatigue cycles applied to a specimen. It is controlled with computer commands and can count up to 16 777 215 cycles.

System Configuration

The photographs in Figs. 1 to 4 show the system configuration. Figure 1 shows the computer system with typical electronics used for control of a fatigue machine. The computer system uses a video monitor for display, a cassette recorder for permanent storage of operating programs, and a thermal printer for hard copy. The fatigue machine control electronics consists of an MTS 436 control unit, 406 controller, and a 430 digital indicator for monitoring maximum and minimum points of a signal.

Figure 2 shows the internal circuit boards of the computer system. Figure 3 shows a rear view of the computer where the inputs and outputs to the control boards are made. There is an input for a ± 10 V external waveform to the function generator board if other than a sine waveform is required for the load waveform. Figure 4 shows the load frame presently used with a tubular-type oven mounted on it for high-temperature crack growth tests.

Specimen Preparation

A test program is currently underway to provide fatigue crack growth rate data for a new titanium alloy subsequently referred to as Type A. The computer control system described has been used for the generation of these data.

The specimen used in the program is of the compact tension (CT) type. Its dimensions and the attachment points of the PD wires and current input wires are shown in Fig. 5. The current is input to the specimen through bolts that have been inserted into undersized holes. The PD wires are welded onto the specimen by an electric spark technique.

FIG. 1—*Computer control system with typical electronics used for fatigue crack growth tests.*

Procedures

PD System Configuration

Figure 6 is a schematic of the setup of the system components. For the test program, a d-c level of 4 A is utilized. At this level, the PD of the CT specimen varies between 0.6 mV at the start of a test to approximately 1.4 mV at completion. To monitor this low reading, the signal is amplified by a gain of 3500 in Channel 1 of the low-level amplifier. The 12-bit A-D system operates over a range of 0 to 10 V and is capable of detecting a change in signal level of 2.44 mV. For the specimen used, this change in signal level corresponds to a detection capability of approximately 25 μm.

The effects of thermal drift and slight changes in the 4 A current level are accounted for by use of the following procedure. The actual current level in the specimen is monitored by passing it through a shunt resistor. At a 4 A level, the shunt voltage is approximately 10 mV. This is amplified by a gain of 500 in Channel 2 of the amplifier. The consecutive readings during the course of testing may then be scaled against the initial current level. This will give a proportional scaling factor which can then be used to correct the PD. Thermal drift is corrected by measuring the PD at a zero level of current and at a

FIG. 2—*Top view of control circuity of computer system.*

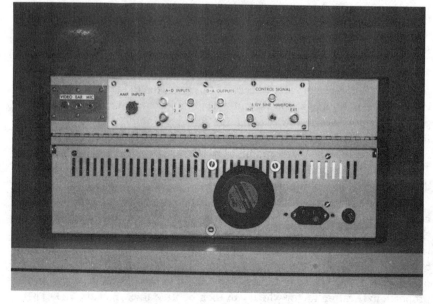

FIG. 3—*Rear view of computer control system.*

FIG. 4—*Load frame and oven used for high-temperature fatigue tests.*

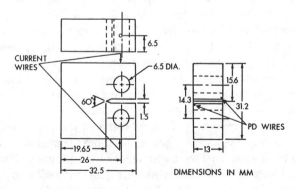

FIG. 5—*Compact tension specimen used in test program.*

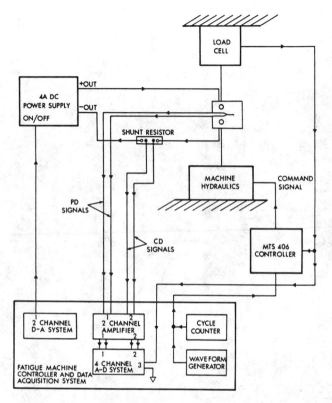

FIG. 6—*Schematic showing interfacing between computer system and fatigue crack growth system.*

4 A level. At zero current, any offsets present are those due to thermal effects. By taking the difference between the voltage at 4 A and zero current, the PD for a given crack length is determined. This is the value that is corrected when a change in current level has occurred. The 4 A supply is switched between zero and 4 A by use of a solid-state relay. The microprocessor system used one of the D-A output signals to provide this control.

A calibration equation relating the crack length to the PD reading was originally obtained from the gas turbine engine manufacturer. This was derived by finite-element methods. The accuracy was checked by the analog foil method and by beach markings of actual compact tension specimens of a commercial titanium alloy, a steel, and a nickel-based superalloy. However, problems began to appear in the early stages of the program. The visually measured crack length was consistently longer than that indicated by the PD.

The rough fracture surfaces of the Type A compact tension specimens appeared to indicate areas of uncracked material and a very irregular crack front. This was probably due to the large grain size, approximately 0.5 mm, of the material. The effect of the uncracked regions would be to lower the resistance of the specimen and result in a shorter PD-determined crack length.

A new calibration equation was determined by fatigue cracking a specimen of a similar titanium alloy, Type B, and monitoring the PD using a digital multimeter. This material was used in order to conserve the Type A specimens. Visual examination of the fracture surfaces of Type B specimens indicated similar features to those of the Type A specimens. Crack growth tests on Alloy B also showed an underestimate in crack length. The specimen was fatigue cracked under decreasing maximum load to maintain an approximately constant crack growth rate. Visual side crack length measurements were taken on both surfaces after a crack growth of 0.025 cm had taken place. The PD was measured with the crack fully open. The test was continued until the specimen fractured.

A check of the fracture surface indicated that a very slight tunneling correction, 0.015 cm, would be necessary. A voltage reference, V_0, at a crack length to specimen width (a/W) ratio was determined at $a/W = 0.244$. The data were then subjected to a least-squares curve fit analysis. The resultant PD formulas based on 54 data points evenly distributed between $a/W = 0.238$ and $a/W = 0.833$ were

$$V/V_0 = 0.6281 + 1.9027(a/W) - 2.1631(a/W)^2 + 2.7702(a/W)^3$$

and

$$a/W = -0.9853 + 1.7316(V/V_0) - 0.5765(V/V_0)^2 + 0.07197(V/V_0)^3$$

where V = PD at a given crack length reading.

The equations were checked by manually performing a threshold test on a Type A specimen. The specimen crack length was measured visually on two surfaces and the average value determined. The computer system was set up to monitor the crack length using the PD method. A comparison between the visual results and the PD measurements is shown in Fig. 7. The accuracy of the equations for longer crack lengths has been confirmed by periodic measurements made during subsequent tests.

The formulas developed based on the visual measurements on the Type B specimen have now been utilized to measure the crack lengths in the Type A material.

Load Control Signal

When the computer control system was being designed, it was desired that a fatigue test be run continuously. This necessitated that a function generator be built into the system and that the fatigue machine control signal be latched (that is, held constant without the computer having to output these values

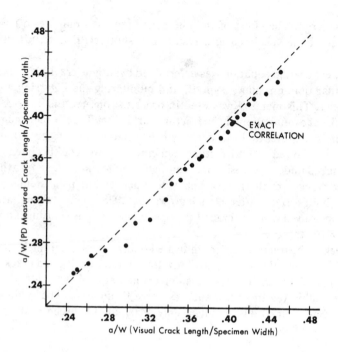

FIG. 7—*Comparison between visual and PD measurement of crack length in a Type A specimen.*

continuously) to allow the computer to monitor other signals without disrupting the fatigue test. For previous systems, the computer could only output the load control signal or measure the crack length. This required occasionally stopping the test while the crack length was being measured.

The diagram in Fig. 8 illustrates how the load control signal is derived. Basically, the load control signal consists of a mean load part and a span control part. The signal is obtained by using the computer to set up two multiplying digital-to-analog converters (DAC's). The input signals to the DAC's are latched. This frees the computer from having to continuously output these values. The span control DAC uses as a reference a ± 10 V sine wave oscillator signal which can be varied between 10 and 80 Hz. The set point or mean load control signal DAC uses a -10 V reference. The two DAC output signals are then added together in a summing amplifier to produce the load control signal.

Software for Fatigue Crack Growth Tests

Two BASIC language programs were developed to perform a complete fatigue crack growth rate test. The threshold test program is used to measure

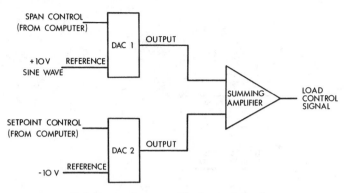

FIG. 8—*Development of load control signal.*

fatigue crack growth rates from the region of 1×10^{-8} m/cycle down to levels of 1×10^{-10} m/cycle. The K-increasing program is used to determine data for tests carried out under constant cyclic load conditions. These programs are discussed separately as follows.

Threshold Test Program

A flow chart describing the threshold test program is shown in Fig. 9. The program utilizes procedures as outlined in the ASTM Test Method for Constant-Load-Amplitude Fatigue Crack Growth Rates Above 10^{-8} m/Cycle (E 647-78T) and the proposed method for measuring ΔK_{th}. The program operates as follows. The specimen dimensions, waveform data, and initial stress-intensity range, ΔK, are input at the start of the program. The maximum, minimum, and mean loads for the required waveform are then computed. At this point, the test is continued or terminated. If the test is continued, the subroutine "LOAD ADJUST" is entered. This outputs a sine waveform for the load control signal and checks that the desired loads are maintained within specified tolerances of 1% for the mean load and 0.5% of the amplitude required. The frequency of this signal is usually 50 Hz. The loads are checked by means of a machine language program that determines the maximum and minimum loads from the load cell. If the loads required are outside the required tolerances, the computer then corrects the voltages to the span and setpoint DAC's as mentioned previously. After this is completed, the crack length during the present crack growth interval is then measured. The loads check and crack measurement subroutines together take approximately 10 to 15 s. This is repeated until the crack has grown more than 0.00625 cm from the previous load shed. To increase accuracy, the crack length measurement used is the average of 20 measurements based on the maximum PD determined over 20 fatigue cycles. Another load shed is per-

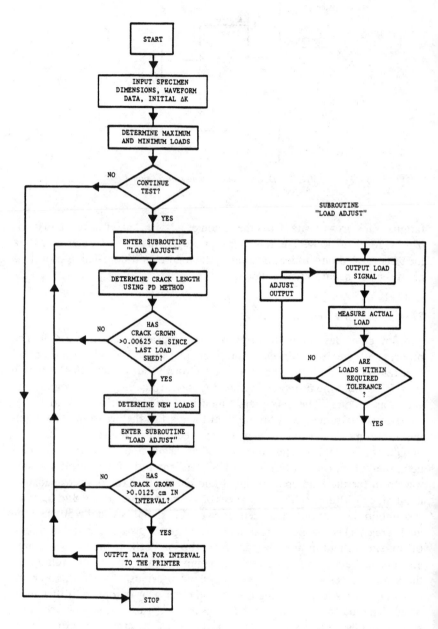

FIG. 9—*Flow chart of threshold test program.*

formed after 0.00625 cm of crack growth has occurred. The amount of load shed resulted in a constant linear decrease in the size of the monotonic plastic zone at the crack tip with increasing crack length. This was accomplished by use of the equation in the proposed ΔK_{th} standard

$$\Delta K = \Delta K_0 \exp \left[C(a - a_0) \right]$$

where

ΔK_0 = initial stress-intensity range at start of test,
a_0 = initial crack length at start of test, and
C = constant with dimensions of 1/length.

The value chosen for C for the test program was -79.74 m^{-1}. This is the value recommended for threshold testing to minimize plasticity effects at the crack tip which may cause premature crack arrest if the load is decreased at too fast a rate. After 0.0125 cm of crack growth has occurred, the crack growth rate over that interval is calculated by use of the secant method. This crack growth interval is well below the maximum of 0.02 W in the ASTM E 647-78T standard. The load shed is made after 0.00625 cm of crack growth to approach a continuous load shed scheme. Data consisting of such information as the average crack size, ΔK, and the crack growth rate over that interval are then output to the printer. The test is allowed to continue until crack growth rate levels near 1×10^{-10} m/cycle have been obtained.

K-Increasing Program

The operation of the K-increasing program was very similar to the threshold program. The major difference was that the loads calculated at the start of the test were maintained constant to test completion. Crack growth rate data were also calculated at intervals of 0.0125 cm and output to the printer. The test was continued until a specified lower number of cycles, usually 5000, was reached for a crack growth interval. At this point the test was stopped and restarted with a lower output frequency waveform, usually 10 Hz, to determine the very fast crack growth rate part of the curve. This lowering of the frequency is necessary because the computer system cannot function properly in this region at high frequencies. Lowering the frequency minimized part of this problem.

Experimental Work

Results

Computer System Operation—The computer system experienced many problems during the developmental process. These involved two broad categories: hardware and software. Hardware problems resulted due to the unique construction of the ZX-81 computer. Certain electronic signals present on the computer did not allow expansion to a bus-based system in a straightforward manner. Inexperience with digital electronics design also prevented the quick solution of certain problems encountered. Software problems mainly involved the development of some of the subroutines used for such purposes as analog-to-digital conversion and for adjustment of the load control signal. The writing of machine language programs proved difficult and modifications were very time-consuming.

The time involved in design, construction, and software development was approximately one year. No breakdown of labor costs can be given because the project was basically a part of a Ph.D. program.

Fatigue Crack Growth Rate Test Results—The program has begun to yield results on the fatigue crack growth behavior of one of the titanium alloys in the test program. The program requires two alloys to be tested under a number of different R ratios and temperature conditions. Figure 10 shows some initial data produced for this program by use of the computer system. The graph shows the results of three fatigue crack growth rate tests performed at $R = 0.60$ and 20°C on the Type A alloy. Each test consisted of a threshold test followed by a K-increasing test. The K-increasing test was begun at a slightly higher load than the final datum point of the threshold test performed on that specimen.

The data of the threshold tests and K-increasing tests appear to match very closely. The data indicated a threshold stress-intensity range, ΔK_{th}, of approximately 4 MPa $\sqrt{\text{m}}$.

Discussion

The operation of the computer system has proved to be quite satisfactory. The initial problems associated with the design and testing of the system have now been solved. The following comments can be made based on the results produced at this time.

The threshold and K-increasing programs are working satisfactorily within the guidelines of the present ASTM standards and recommended practices related to fatigue crack growth rate testing. The system appears to be capable of producing consistent data with crack growth rate intervals of a length one-quarter the size in the standard. In Fig. 10, for clarity, all the data have not been shown. However, using the smaller crack growth interval resulted in over 100 datum points per test.

The load-shedding scheme utilized does not result in any retardation in crack growth rate when testing under laboratory air conditions. According to recommended ΔK_{th} test practice, the load shed used would be referred to as continuous load shedding. The data in Fig. 10 indicate no apparent mismatch between the data of the threshold and K-increasing portions of the

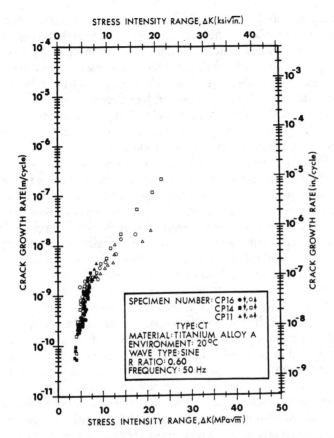

FIG. 10—*Fatigue crack growth rate test results at* R = 0.60 *and* 20°C.

test. The value of C chosen was probably conservative. A value of C allowing a faster load shed would probably be acceptable without causing retardation of crack growth. Such an effect needs to be the object of further consideration as test time is greatly influenced by the value of C chosen.

The early problems encountered in regard to disagreement between visual and PD measurements of the specimen crack length were probably due to the resultant crack growth morphology of the titanium alloy. The large grain size of the material and the cracking mode resulted in a very rough fracture surface. Crack closure effects [3] probably resulted in a recontacting between fracture surfaces, thus causing a change in the PD readout. The effects of crack closure are more significant at low R ratios, long crack lengths, and low loads. In the threshold region, the effects due to crack closure would become more pronounced.

The disagreement between visual and PD measured crack length was solved by obtaining an actual PD-versus-a curve from a sample of titanium alloy under fatigue loading conditions. The problems encountered in performing this test are due to the unevenness of crack growth through the specimen for a large-grained material. If the crack is growing around a favorably oriented grain on the side of the specimen, the crack as measured visually would appear to be much longer than the average crack length. In subsequent tests after the new calibration equation had been obtained, a check of the crack length visually on one surface would often show a substantial disagreement with the PD measured crack length. Only after final fracture could the accuracy of the PD measurements be checked.

The problems experienced with the PD calibration of the system emphasized the need to establish a standard method for approaching the setup and calibration of such a system. The same would be necessary for using crack opening displacement (COD) techniques. Experience with both methods indicated that one method may be preferred over the other depending on such factors as material type to be tested and the extrinsic test parameters. The overall accuracy and reliability of these methods must also be further quantified. For example, the method developed by Hewitt [4] for evaluating the COD method needs to be further investigated. Finally, a standardized method of correcting, if necessary, the crack growth data produced by such automated methods by using visual measurements of the fracture surface is required. The inclusion of information such as given by Saxena [5] on data-correcting techniques is seldom seen. Often the reader is given the impression that the test technique used by an author is perfect and results in data that need no correction. This seldom proves to be the actual case.

The exact form of a test standard for using PD methods of measuring crack length would probably not contain a universal PD equation for each test specimen design currently used in crack growth testing; however, the procedures used to establish such a curve may be standardized. A suggested PD setup for each specimen design and a PD calibration equation may be included in such a standard together with required precautions regarding use. Also, for the successful implementation of a PD system, it would be necessary to provide information on minimizing electrical noise and thermal effects from interfering with the PD signal. This information would take the form of recommended wiring procedures for carrying low-level signals from the test specimen to the monitoring equipment and guidance for establishing the necessary environmental controls around the test equipment to minimize effects on crack length measurements.

Conclusions

The design and construction of a computer system for use in fatigue crack growth rate testing have been described. The successful operation of the system indicated that such equipment can be used to cut down on the time and expense involved in acquiring such data. The problems encountered incorporating the PD method of measuring crack length emphasized that work must still be completed on standardizing test procedures that utilize remote crack length monitoring techniques. Further investigations also must be conducted to better understand such effects as retardation of crack growth which limit how fast ΔK_{th} may be obtained during a test.

Acknowledgments

The authors wish to acknowledge the Rolls Royce Aero Engine Division for providing financial support of the work, the University of Toronto, Department of Mechanical Engineering, for use of facilities, and MTS Systems for providing a fellowship for P. M. Sooley's Ph.D. program.

References

[1] Dickson, J. I., Baïlon, J. P., and Masounave, J., "A Review on the Threshold Stress Intensity Range for Fatigue Crack Propagation," *Canadian Metallurgical Quarterly*, Vol. 20, No. 3, 1981, pp. 317-329.
[2] Bucci, R. J., "Development of a Proposed Standard Practice for Near-Threshold Fatigue Crack Growth Rate Measurement," ALCOA Report No. 57-79-14, Aluminum Company of America, Pittsburgh, 1979.
[3] Elber, W., "Fatigue Crack Growth Under Cyclic Tension," *Engineering Fracture Mechanics*, Vol. 2, 1970, pp. 37-45.
[4] Hewitt, R. L., "Accuracy and Precision of Crack Length Measurements Using a Compliance Technique," *Journal of Testing and Evaluation*, Vol. 11, No. 2, 1983, pp. 150-155.
[5] Saxena, A., "Electrical Potential Technique for Monitoring Subcritical Crack Growth at Elevated Temperatures," *Engineering Fracture Mechanics*, Vol. 13, 1980, pp. 741-750.

Yoshiyuki Kondo[1] and Tadayoshi Endo[1]

An Automatic Fatigue Crack Monitoring System and Its Application to Corrosion Fatigue

REFERENCE: Kondo, Y. and Endo, T., "An Automatic Fatigue Crack Monitoring System and Its Application to Corrosion Fatigue," *Automated Test Methods for Fracture and Fatigue Crack Growth, ASTM STP 877,* W. H. Cullen, R. W. Landgraf, L. R. Kaisand, and J. H. Underwood, Eds., American Society for Testing and Materials, Philadelphia, 1985, pp. 118–131.

ABSTRACT: A measuring system that automatically detects fatigue crack initiation and propagation behavior has been developed. This system is based on calculating and recording automatically the change of compliance due to fatigue crack growth using an analog circuit.

The authors applied this system to fatigue testing in an aqueous environment and succeeded in detecting a part-through semi-elliptical crack as small as 0.2 mm deep initiating from the notch root.

Using this system, the crack initiation and following propagation behavior in the vicinity of a notch root in 3.5Ni-Cr-Mo-V steel in an aqueous environment were investigated.

KEY WORDS: corrosion fatigue, crack initiation, crack growth, compliance, crack closure

In the case of corrosion fatigue that is often encountered in industrial machinery, the crack initiates at a fairly early stage, and crack propagation occupies the major part of the fatigue life. In such circumstances, it is important to know both the fatigue crack initiation and the following crack propagation characteristics of a short crack for the purpose of determining the inspection interval and so on.

It is not always possible, however, to detect the crack initiation characteristic separately; that is, commonly used optical monitoring is difficult for a specimen enclosed in an autoclave. Moreover, continuous monitoring by the human eye is almost impossible. Therefore, in order to obtain such data, an

[1]Research engineer and manager, respectively, Material Strength Research Laboratory, Takasago Technical Institute, Mitsubishi Heavy Industries, Ltd., Takasago, Japan.

118

automatic measuring system to continuously monitor the fatigue crack length is desirable.

There are many methods which can be applied to automatic measurement; for instance, crack gage, electric potential, eddy current and compliance methods. Among these, the compliance method has excellent characteristics compared with the others, such as its applicability in various environmental conditions, its resistance to electrochemical disturbance during the corrosive process, and its insensitivity to temperature drift.

This paper reports the development of an analog circuit to calculate and record automatically the change of compliance due to fatigue crack initiation and the crack propagation that follows.

Automatic Measuring System

This system, which uses the compliance method for crack detection, is composed of two analog circuits as shown in Fig. 1.

The first is the subtraction circuit of the "unloading elastic compliance method" [1]. Consider a case where a fatigue crack initiates and grows from a notch root. As shown schematically in Fig. 1b, the load-displacement curve inclines as indicated by a_1, a_2, . . . as the crack grows. In the conventional compliance method, the gradient of load-displacement curve ($d\delta/dP$) is converted into crack length. This method as it is, however, does not exhibit a high enough sensitivity for detecting fatigue crack initiation because the change of gradient due to crack growth is very small.

In this report, therefore, we utilize the subtraction circuit of the "unloading elastic compliance method" for the purpose of improving the accuracy of crack detection. Load (P) and displacement (δ) signals are put into the subtraction circuit. In this circuit the initial displacement δ_0 (without crack) is subtracted from the actual displacement δ (with crack), which results in the subtracted displacement $\delta_{sub} = \delta - \delta_0$. As shown in Fig. 1c, the subtracted load-displacement curve ($P \sim \delta_{sub}$) will be parallel to the load axis indicated as a_0 for the initial state and then will begin to incline as the crack initiates and grows. The gradient of the subtracted load-displacement curve ($d\delta_{sub}/dP$) corresponds to the propagated crack length. Since δ_{sub} does not contain the elastic component which usually occupies the major portion of the signal, it can be magnified to any arbitrary degree we want. Thus the change of gradient due to crack growth becomes much larger compared with the conventional compliance method and hence the accuracy of crack detection has been improved.

The second circuit is to calculate the gradient ($d\delta_{sub}/dP$) automatically. For this purpose we have to overcome two problems. The first is that there often exists some drift in the signal that is caused mainly by temperature drift of the environment. So, if we plan to obtain the gradient by simply dividing δ_{sub} by P, some error will arise. Hence we have adopted the differentiators to

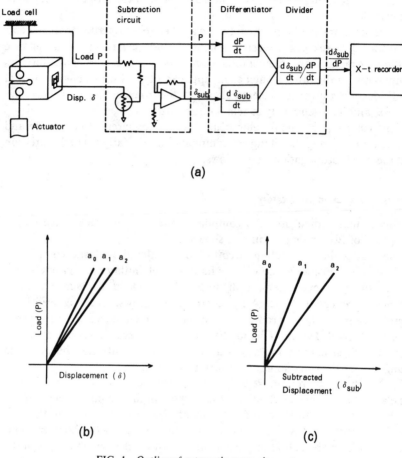

FIG. 1—*Outline of automatic measuring system.*

eliminate the influence of drift in signal. As shown in Fig. 1a, the signals P and δ_{sub} are differentiated by time (t), and we obtain dP/dt and $d\delta_{sub}/dt$, respectively. The influence of the drift, which occurs at a fairly lower frequency than test frequency, was completely eliminated. Then $d\delta_{sub}/dt$ is divided by dP/dt by an analog divider. Here we obtain $d\delta_{sub}/dP$. The actual circuit diagram is shown in Fig. 2. The analog differentiator is an operational amplifier.

The second problem is to eliminate the influence of crack closure [2]. As shown in Fig. 3, the actual load-displacement curve of a cracked specimen bends because of the change of compliance due to the crack closure phenomenon, and only the gradient for the region above the closure level corresponds to the crack length. Therefore $d\delta_{sub}/dP$ should be output only for this por-

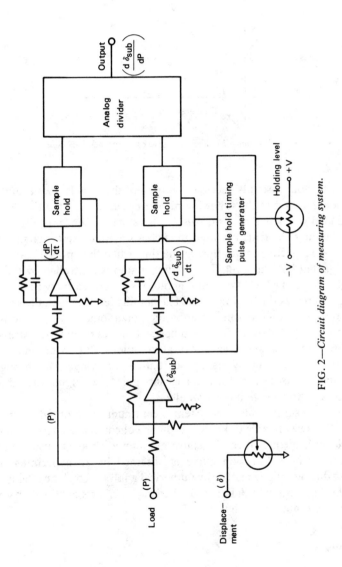

FIG. 2—*Circuit diagram of measuring system.*

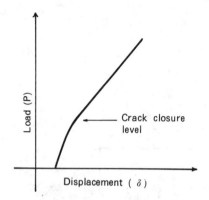

FIG. 3—*Load-displacement curve of a cracked specimen.*

tion. For this purpose it is necessary to prevent the signals (dP/dt and $d\delta_{sub}/dt$) from entering into the following divider circuit during the time when the applied load is below the closure level. Therefore, as shown in Fig. 2, signal sample holding elements are used in such a way that these signals are held constant at respective values during the time when the applied load is below the designated holding level. The holding action is controlled by the timing pulse generator in which a controlling pulse is generated every time when the load signal crosses the holding level. The holding level is manually set slightly higher than the crack closure level by the potentiometer. The holding level relates to the crack closure point and hence it relates to material, stress ratio, stress-intensity level, and environmental conditions. But, just as in this study, it will be sufficient in many cases to set it at the mean load level throughout a test. In special cases where the crack closure point rises higher than this level, it is necessary to change it occasionally.

The voltage signal from this system is continuously recorded by an X-t recorder. Thus we are able to know the change of compliance against the loading cycle and, further, it can be converted into crack length using a calibration curve. The calibration curve is obtained by an experiment (beach marking during fatigue test) or by numerical analysis, etc. Thus the measurement of fatigue crack initiation and following crack propagation behavior can be carried out automatically.

Procedure

Testing Apparatus

An electrohydraulic fatigue testing machine (capacity ±98 kN) was used. For the testing in an aqueous environment, an environmental cell was attached as shown in Fig. 4a. Test frequency was set at 37 Hz. A general view of the proposed measuring system is shown in Fig. 4b.

Specimen

The type of specimen used is a blunt-notched compact-type (BNCT) specimen as shown in Fig. 5 (dimensions are in millimetres). It has a hole (2.5 mm in diameter) at the tip of the slit. The maximum stress that occurs at the notch

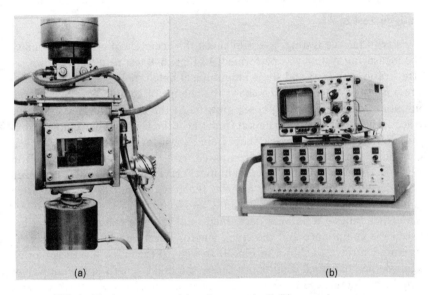

(a)   (b)

FIG. 4—*Testing apparatus:* (a) *environmental cell,* (b) *measuring system.*

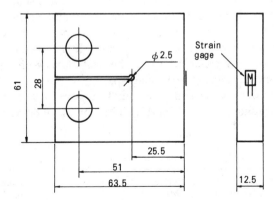

FIG. 5—*Blunt-notch compact-type specimen.*

root was analyzed by the finite-element method (FEM). The inner surface of this hole was finished by a reamer and polished with No. 400 emery paper. The material investigated was 3.5Ni-Cr-Mo-V forged steel. The mechanical properties of this material are listed in Table 1. A foil strain gage 3 mm long was bonded at the center of the specimen thickness at the opposite side of fatigue crack as shown in Fig. 5 and thus the so-called "back face strain" was measured. For the aqueous environment test, this strain gage was coated with waterproofing wax.

Fatigue Test in Air

Prior to fatigue testing, a test to check the accuracy of crack detection of the measuring system was performed. In Figs. 6–8 test results are shown for stress ratio $R = 0.05$ and stress amplitude at notch root $\sigma_a = 337$ MPa. Figure 6 is a part of chart which shows the change of compliance due to crack initiation and the following crack growth. At the point indicated as A, the trace of the record begins to depart from the initial level and detectable cracks initiate here. In Fig. 7 crack shapes are shown for the point of time when compliance increased by 0.2% compared with the initial value (Point B in Fig. 6). Three part-through semi-elliptical cracks as small as 0.3 mm deep were detected. Similarly, Fig. 8 shows the crack shapes at the point when compliance increased by 1.5% (Point C in Fig. 6).

Further fatigue tests were performed to obtain the calibration curve between the compliance and the crack length. Beach marks were left on the fractured surface as shown in Fig. 9. In Fig. 10 the crack length is plotted against the change of compliance (here, the crack length is the equivalent length which gives the same area as the actual crack when the crack shape is assumed to be rectangular). This curve is made from four duplicate tests in order to check the reproducibility. All data are in good agreement. Even in the case of the small part-through crack where the crack shape, the number of cracks, and the sites of crack initiation vary from one specimen to the next, the relation between the equivalent crack length (that is, cracked area) and the change of compliance is nearly the same [3]. It is necessary to recalibrate for different types of specimens because the relation between the compliance and crack length depends on specimen type.

Fatigue tests were performed on nominal stress ratios $R = 0.05$ and 0.8. (Test results are shown later in Fig. 12.) Initiation life of detectable cracks (a

TABLE 1—*Mechanical properties of material.*

0.2% Proof Stress, MPa	Ultimate Tensile Strength, MPa	Elongation, %	Reduction of Area, %
722	838	29.0	74.2

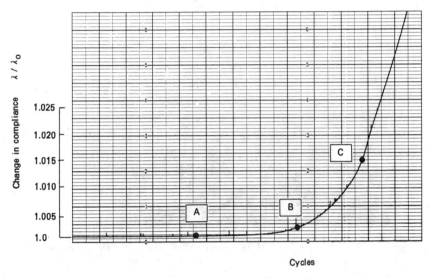

FIG. 6—*Change in compliance due to crack initiation and the following crack growth.*

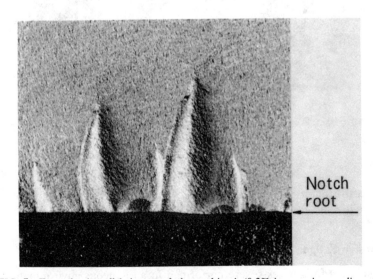

FIG. 7—*Example of small fatigue crack detected in air (0.2% increase in compliance).*

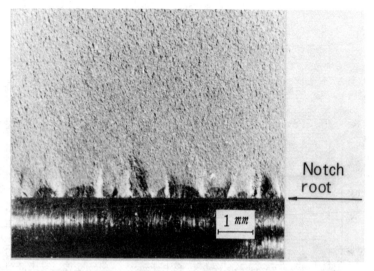

FIG. 8—*Example of small fatigue crack detected in air (1.5% increase in compliance).*

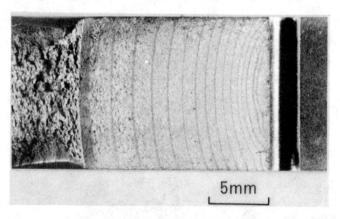

FIG. 9—*Beach marks on fractured surface.*

few part-through cracks as small as 0.2 mm deep) occupies about 50% of fatigue life.

Corrosion Fatigue Test in Aqueous Environment

Corrosion fatigue tests were performed in an aqueous environment. The environmental conditions were: 5% sodium chloride soluted in deionized water, temperature 90°C, $FeCl_3$ added, pH 4.5 ~ 5.0, flow rate 1 L/min.

The accuracy of crack detection was examined also in the aqueous environ-

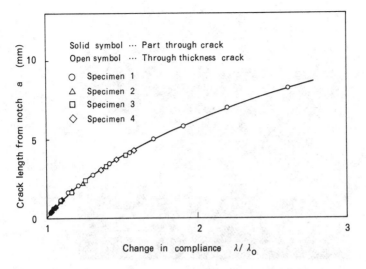

FIG. 10—*Calibration curve between compliance and crack length.*

ment. As shown in Fig. 11, a part-through crack as small as 0.2 ~ 0.3 mm deep was detected.

S-N *Curve of Crack Initiation*

Two different nominal stress ratios, $R = 0.05$ and 0.8, were chosen for the fatigue tests. In this study, we investigated separately the crack initiation and the propagation processes, and the following behaviors were found. Test results are shown in Fig. 12. Open symbols, half-solid symbols, and solid symbols correspond to crack lengths of 0.2, 2, and 8 mm, respectively. In Fig. 12a it is remarkable that the fatigue crack initiation life was less than 10% of fracture life in the long-life region. Moreover, the specimen which did not fracture after 10^8 cycles had a crack that was growing continuously. The *S-N* curve of open symbols indicates that the crack initiation characteristic goes down monotonically. On the contrary, however, a cusp is seen in the *S-N* curve of half-solid or solid symbols, which indicates the fracture characteristic, namely, that the crack propagation life becomes relatively long at intermediate stress levels [4,5].

The cusp in the *S-N* curve is supposed to be caused by stagnation in crack propagation. Since the proposed system can also measure the crack closure phenomenon, the change in crack closure level was monitored. Typical load-displacement curves are inserted in Fig. 12a. At intermediate stress levels only, a remarkable crack closure phenomenon develops, probably because of the corrosive wedging effect, resulting in a decreased effective stress-intensity range. Hence, the crack propagation rate decreases and the fracture life in-

FIG. 11—*Example of small fatigue crack detected in corrosive environment (0.5% increase in compliance).*

creases at intermediate stress levels. As shown in Fig. 12*b* for the case where stress ratio is high, crack closure does not appear and hence stagnation in crack propagation does not occur and the *S-N* curve goes down monotonically.

Crack Propagation Behavior of Short Crack

The crack propagation rate (da/dN) of a growing crack from the notch root for stress ratio $R = 0.05$ was measured and is plotted against ΔK in Fig. 13. The solid line is the crack propagation rate obtained in air at room temperature by the conventional *K*-decreasing method. It is seen that in the aqueous environment, fatigue crack propagation occurs even below the threshold stress-intensity range (ΔK_{th}) obtained in air.

Comparison Between K-increasing and K-decreasing Test Method

A conventional *K*-decreasing test was performed to obtain the ΔK_{th} in a corrosive environment. The result is shown in Fig. 14. The solid symbols indicate the crack propagation rate in the *K*-decreasing test, and open symbols indicate the crack growth rate from the notch. It is seen that crack propagation does occur at an early stage of crack growth from the notch root (open symbols) for ΔK below the ΔK_{th} obtained by the conventional *K*-decreasing method (solid symbols) in the same environment. This is caused mainly by the difference in crack closure as typical load-displacement curves are shown in

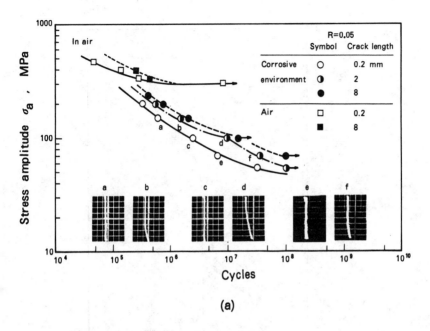

(a)

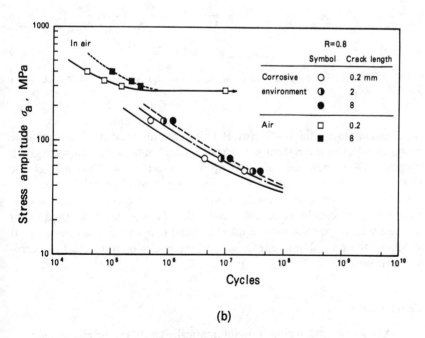

(b)

FIG. 12—*Fatigue test results of 3.5Ni-Cr-Mo-V steel:* (a) *stress ratio* R = *0.05,* (b) *stress ratio* R = *0.8.*

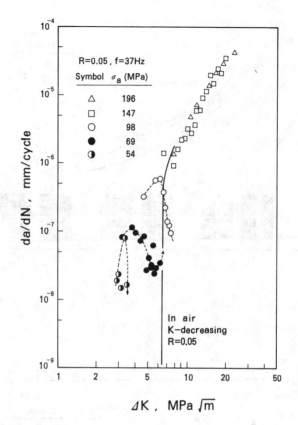

FIG. 13—*Fatigue crack growth from notch root in corrosive environment on 3.5Ni-Cr-Mo-V steel (R = 0.05).*

the figure; namely, the crack closure point goes up because of the corrosive wedging effect in the K-decreasing test, but in contrast with this the crack closure phenomenon does not develop for a short crack. Thus it might be unsafe if we use ΔK_{th} obtained in the conventional K-decreasing test as a criterion for a short crack growing from the notch in a corrosive environment. Hence it can be concluded that the conventional K-decreasing method might not be a suitable method when applied to a test in a corrosive environment. It is desired to establish an appropriate test method to obtain ΔK_{th} in a corrosive environment.

Conclusions

1. A low-cost analog circuit to automatically calculate the change of compliance due to crack growth with high sensitivity has been developed. This system was applied to the corrosion fatigue test in an aqueous environment.

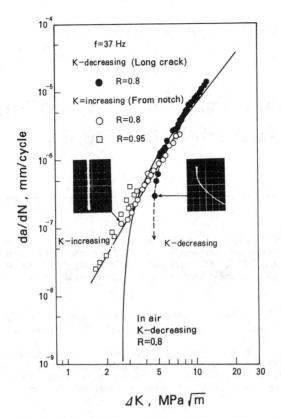

FIG. 14—*Effect of test condition on fatigue crack propagation rate in corrosive environment.*

2. Fatigue cracks as small as 0.2 mm deep initiating from the notch root could be detected.

3. Fatigue crack propagation occurred at an early stage of crack growth from the notch root for ΔK below ΔK_{th} obtained by the conventional K-decreasing method.

References

[1] Kikukawa, M., Jono, M. and Tanaka, K. in *Proceedings*, Second International Conference on the Mechanical Behavior of Materials (ICM II), Boston, 1976, p. 254.
[2] Elber, W. in *Damage Tolerance in Aircraft Structures, ASTM STP 486*, American Society for Testing and Materials, Philadelphia, 1971, p. 230.
[3] Song, J. and Heckel, K. in *Fracture Mechanics in Engineering Application*, G. C. Shih and S. R. Valluri, Eds., Sijhthoff and Noordhoff, Alphen aan den Rijn, The Netherlands, 1979, p. 557.
[4] Iwamoto, K., *Transactions*, Japan Society of Mechanical Engineers, Vol. 30, No. 212, 1964, p. 500.
[5] Endo, K., Komai, K., and Kinoshita, S. in *Proceedings*, 22nd Japan Congress on Materials Research, Kyoto, Japan, 1978, p. 193.

W. Alan Van Der Sluys[1] and Robert J. Futato[1]

Experience with Automated Fatigue Crack Growth Experiments

REFERENCE: Van Der Sluys, W. A. and Futato, R. J., **"Experience with Automated Fatigue Crack Growth Experiments,"** *Automated Test Methods for Fracture and Fatigue Crack Growth, ASTM STP 877*, W. H. Cullen, R. W. Landgraf, L. R. Kaisand, and J. H. Underwood, Eds., American Society for Testing and Materials, Philadelphia, 1985, pp. 132–147.

ABSTRACT: The experience of the authors with the automation of the fatigue crack growth experiment is described. The system is described from both the hardware and the software standpoints.

The approach used in automating the fatigue crack growth experiment was to make as much use of the computer as possible by writing large, user friendly, multipurpose program systems. The approach has been very successful based on four years of experience in conducting experiments. High quality results have been obtained from experiments conducted by personnel not skilled in computer programming.

KEY WORDS: automatic test control, computerized data acquisition, fatigue crack growth, pressurized water reactor environments, constant stress-intensity factor (ΔK), SA-533B steel

The fatigue crack growth experiment is a prime candidate for automation. The experiment is tedious, requires continuous monitoring, and must be performed with a high degree of accuracy. In addition, some fatigue crack growth experiments such as threshold testing and low-frequency corrosion fatigue experiments can last for weeks and even months. The automation of the fatigue crack growth experiment, therefore, offers the possibility of a large return on the initial investment both in terms of the reduction in man-hours to conduct the experiment and in terms of accuracy of the results.

The objective of this paper is to discuss the authors' experience with the automation and performance of fatigue crack growth experiments. The automated test system is described from both the hardware and software standpoints. Possible modifications to the ASTM Test Method for Constant-Load-

[1]Technical advisor and research engineer, respectively, Babcock & Wilcox, a McDermott company, Alliance Research Center, Alliance, OH 44601.

Amplitude Fatigue Crack Growth Rates Above 10^{-8} m/Cycle (E 647-83), are proposed to make the standard more compatible with automated testing methods.

Approach to Automation

The approach used in automating the fatigue crack growth experiment is to make as much use of the computer as possible by writing large, user-friendly multipurpose program systems. In this way experiments can be run by people skilled in materials testing but not necessarily skilled in computer programming.

At this time the authors have two automated testing systems in which fatigue crack growth experiments can be conducted. In one system, three MTS servohydraulic testing machines are interfaced to a PDP 11/34 computer through an MTS 433 Processor Interface Unit. In the second system, four MTS servohydraulic actuators are installed on autoclaves, allowing fatigue crack growth tests to be conducted in high-temperature, high-pressure water environments. These four loading systems are also interfaced to a single PDP 11/34 computer through an MTS 433 Processor Interface Unit. Each of the PDP 11/34 computers has two hard disk drives and two flexible disk drives interfaced for data storage. Figure 1 presents a schematic of the computer control system. A more detailed description of the system can be found in Refs 1 and 2.

In the fatigue crack growth experiments, the specimen compliance is used to determine the crack length [3]. In order to make the compliance measurement with a high degree of confidence using the 12-bit analog-to-digital (A/

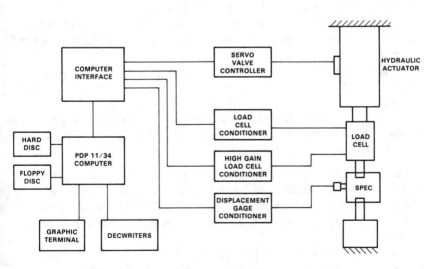

FIG. 1—*Schematic of computer control system.*

D) converters available in the MTS computer interface, a crack length measurement system independent of the specimen load control system is used. This system consists of two amplifiers, a displacement gage for monitoring the opening displacement on the specimen, and a dual-bridge load cell. One bridge of the load cell is used by the servo control system to measure and control the load on the specimen while the second bridge is used in the crack length measuring system. The amplifiers in this system are operated at a relatively high gain so that the amplitude of the signal produced during a fatigue cycle will be close to the full-scale output of the amplifier. In order to keep the signals on scale as the test progresses, a digital-to-analog (D/A) converter in the computer interface is used to supply a bias voltage to the amplifier which in effect rezeros the amplifier as the test progresses. Since the only information needed from this crack length measurement system is the specimen compliance, which is not a function of the absolute value of the load or displacement but rather only the relative changes in these values, the system just described produces a measure of the compliance, and hence the crack length, to a high resolution.

The computer program which controls the fatigue crack growth experiments is written in MTS Systems Corp. Multi-User Basic language such that three different types of fatigue crack growth experiments can be conducted. Conventional constant load experiments, constant-ΔK experiments (in which ΔK is held constant during the test), and decreasing-ΔK experiments can all be performed using the test control software.

The fatigue crack growth test program is divided into four separate programs:

1. *Input Stage*—This program asks the operator for all the information required to describe the specimen and test conditions.

2. *System Checkout*—This program measures the initial crack length and tests to see if the hydraulics are able to meet the test conditions (loads and frequency).

3. ΔK *Control Program*—This program runs the test. It acquires and stores crack growth data and adjusts the applied load to yield the specified ΔK level if constant-ΔK or decreasing-ΔK tests are desired.

4. *Data Analysis*—This program performs data analysis. The crack length versus cycles information is displayed and crack growth rate versus applied ΔK determined.

The input stage program must be run first since it asks the operator for information which describes each test to be performed. This information includes the type of test to be run, the specimen type and material, the desired initial ΔK for the test, the initial and final crack lengths, the transducer full-scale ranges, etc.

Certain calculations are performed to aid the user in answering the input questions. For instance, using the input initial and final crack length and initial and final ΔK-values, the program calculates the maximum load and displacement expected during the test so that the transducer ranges can be chosen appropriately.

A table of all the test parameters that were input and a summary of the maximum and minimum loads and displacements are printed. The program then allows the operator to change any parameter without reentering all the input parameters.

When the operator is satisfied with all the input test parameters, the parameters are written to a disk file for later use. The program then chains into the system checkout program.

The main function of the system checkout program is to perform some initial cycling of the specimen to check for any obvious problems with the test setup. The program starts by reading the parameter file and printing out a series of statements instructing the operator to perform certain tasks. These tasks include starting the hydraulic power supply, ensuring that the analog controller is set to the computer-controlled mode, and setting the transducer amplifiers to the proper ranges.

The initial crack length in the specimen is then measured. This is done by calculating a load based on the estimated initial crack length and an initial cycling ΔK. This load is then applied to the specimen and the specimen compliance is measured. The crack length is calculated and printed out along with the initial crack length that was input. At this point, if the operator is satisfied with the crack length measurement, he can run the test. If he is not satisfied, he can take action to correct the problem. If the operator chooses to run the test, the program chains into the ΔK control program which actively controls the running of the test.

As the test is running, the crack length is determined by measuring the specimen compliance at increments in cycles specified by the operator. An average of 80 load and displacement points is taken during an unloading half-cycle, and the slope of the unloading line calculated using a linear regression analysis of the data. This is repeated on three consecutive unloadings to allow an average crack length to be determined. Each of the three crack length values and the 95% confidence interval for each value are displayed along with the average crack length and the cycle count.

Depending on which type of test is being conducted, after the crack length measurement the loading of the specimen is either continued at the previous level or a new load is determined based on the new crack length and the desired ΔK level.

The test continues until the specimen has failed, a preset crack length is reached, or a preset ΔK level is reached. When a test is completed, the program is able to sequence into another test without operator interruption. In this way a series of experiments can be conducted on the same specimen without operator action.

The data analysis capabilities of the system allow for two different types of analysis, depending on the type of test being conducted. For constant-ΔK experiments where the a-versus-N curves are straight lines, a linear regression analysis is used to determine the crack growth rate. For constant-load experiments a modified ASTM E 647 secant procedure is used to determine the crack growth rate. This is discussed in more detail later.

Test Experience

All three types of tests which the program is capable of performing have been conducted. The vast majority of the test work has been constant-ΔK testing in autoclave environments, although a limited amount of constant-load testing has also been performed. Threshold tests have not been performed using the ΔK shedding feature of the program, but the ΔK shedding feature has been used to reduce the ΔK level or the R-ratio from one test condition to the next on the same specimen.

Figure 2 is an example of the type of results obtained from a constant-ΔK fatigue crack growth experiment. The data presented in this figure were obtained in 288°C water on SA-533 steel at ΔK ranges of 33 and 44 MPa$\sqrt{m}$ and an R-ratio of 0.5. The crack growth rates presented in this figure were

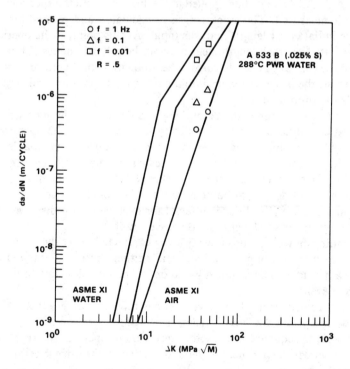

FIG. 2—*Fatigue crack growth in a 288°C pressurized water reactor (PWR) environment as a function of ΔK and loading frequency.*

obtained from the crack length versus cycles information shown in Figs. 3 and 4. A linear regression analysis was used to determine the slopes of these lines. These figures represent an ideal type of result. The crack growth rate lines are linear and there is very little scatter in the crack length information. The confidence interval obtained from the linear fit is also presented on the figures. There is a high degree of confidence in the crack growth rate determined from this type of information. The 95% confidence band on each of these six crack growth rates is between one and three orders of magnitude smaller than the measured crack growth rate. This means that for the crack growth increment over which the experiment was performed the crack growth rate was determined to a relatively high degree of confidence.

Figures 5 and 6 represent the results of a load shedding experiment in which ΔK was dropped from 25 to 12 MPa$\sqrt{m}$ over a crack length increment of 0.10 mm. In Fig. 5 the crack length versus cycles information obtained during this experiment is presented, while the crack length versus applied ΔK information is presented in Fig. 6. The load shedding procedure makes use of the following expression found in Ref 4 for gradually dropping the applied ΔK level

$$\Delta K = \Delta K_0 \exp\left[C(a - a_0)\right]$$

where

$$\begin{aligned}
a_0 &= \text{initial crack length at beginning of } \Delta K \text{ shed,} \\
a &= \text{instantaneous crack length,} \\
\Delta K_0 &= \text{alternating stress-intensity factor at crack length } a_0, \\
\Delta K &= \text{alternating stress-intensity factor at crack length } a, \text{ and} \\
C &= \text{decay constant.}
\end{aligned}$$

For the data shown in Figs. 5 and 6, the decay constant, C, was -2. In general, it has been found that decay constants up to -7 will work satisfactorily. Therefore, when it is desired to drop the applied ΔK level, the crack length increment over which the applied ΔK is to be dropped is determined and the resultant decay constant is calculated. If the decay constant is found to be greater than -7, the experiment is run; if it is less than -7, the crack length increment is increased and a new decay constant is calculated.

The experience using this approach has been very good. The data presented in these figures are typical of most tests conducted.

Figures 7 and 8 present the results of a constant-load test conducted in 288°C water on an SA-533 material with a sine wave loading frequency of 0.017 Hz. Figure 7 presents the crack length versus cycles information obtained during this experiment while Fig. 8 presents the da/dN-versus-ΔK information. The three solid lines in this figure represent the reference flaw growth curves found in Section XI of the American Society of Mechanical Engineers Boiler and Pressure Vessel Code for air, and for low and high R-

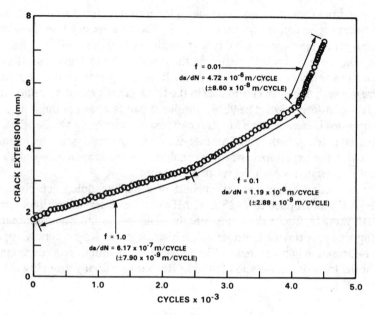

FIG. 3—*Crack length versus cycles information for three experiments conducted at a ΔK level of 44 MPa$\sqrt{m}$ on SA533B-1 material in a 288°C PWR water environment.*

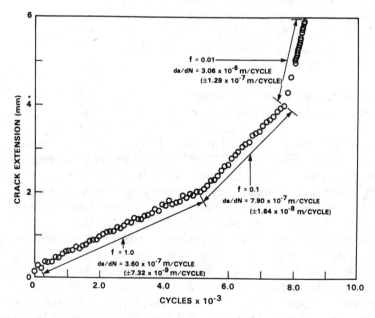

FIG. 4—*Crack length versus cycles information for three experiments conducted at a ΔK level of 33 MPa$\sqrt{m}$ on SA533B-1 material in a 288°C PWR water environment.*

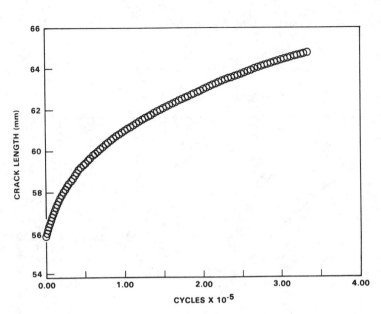

FIG. 5—*Crack length versus cycles information obtained from a* ΔK *shedding experiment on SA533B-1 material in a 288°C PWR water environment.*

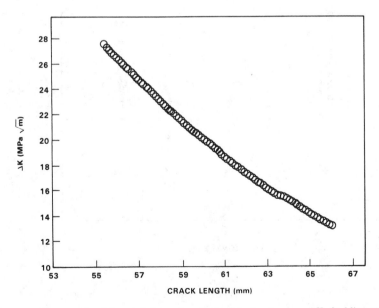

FIG. 6—ΔK *versus crack length information obtained during the same* ΔK *shedding experiment reported in Fig. 5.*

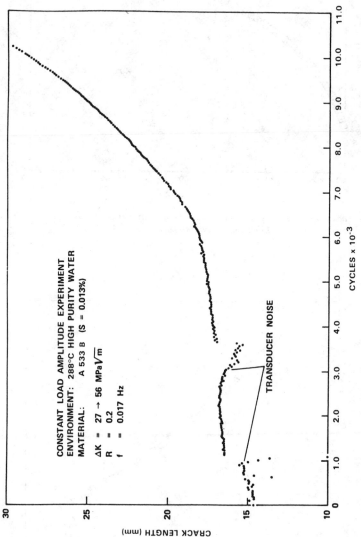

FIG. 7—*Crack length versus cycles information obtained during a constant-load fatigue crack growth experiment performed on SA533B-1 material in 288°C high-purity water.*

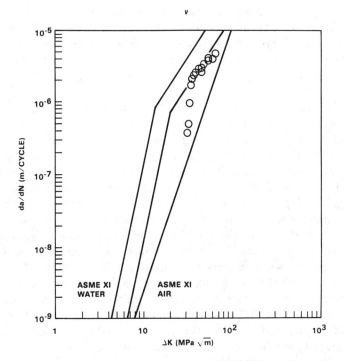

FIG. 8—*Cyclic crack growth rates determined from the crack length information presented in Fig. 7.*

ratios in a high-temperature water environment. The crack growth rate data plotted in this figure were obtained from the crack length versus cycles data presented in Fig. 7 using a modified secant method described later.

Discussion of Test Results

It is difficult to analyze the test results just presented in accordance with the ASTM E 647 standard. As in most automated testing procedures, it is the total crack length and not the change in crack length that is measured (in contrast to many manual techniques). Many more crack length measurements are made per extension interval than are required by E 647. The E 647 procedure recommends that the crack length be measured at a constant increment of 0.25 mm or ten times the crack length measurement precision, whichever is greater. This crack length versus cycles information is then analyzed using either the secant method or the incremental polynomial method to obtain the crack growth rate, da/dN, versus ΔK, information.

In order to meet these requirements, the investigator with an automated facility must develop a procedure to obtain crack length versus cycles information at constant increments of crack length rather than at constant increments of cycles, which is more conventional with automated equipment. This means that crack length information is obtained on increments of cycles and discarded until the required crack length increment is reached, with all the crack length information obtained between the Δa increments being disregarded.

This procedure seems inappropriate for three reasons. First, the crack length data which are discarded contain information about the random noise or statistical scatter in the crack length measurement technique being used, even if no physical crack extension has occurred between adjacent data points. Such information, when available, should not be ignored but should be used to minimize the effect of noise on the calculated crack growth rate. Secondly, by using the first crack length reading which exceeds the required growth increment to determine the cycles information, only the crack length information on the upper bound of the scatterband in crack length is being used in the rate calculation. This reduces the accuracy in the determination of the cyclic crack growth rate. Finally, the use of multiple crack length readings within the prescribed E 647 limits helps reduce the sensitivity of the calculated crack growth rate to local variations in properties. By using numerous, closely spaced data points in the determination of one crack growth rate, the chance of a small number of locally deviant data unduly influencing the calculated crack growth rate is virtually eliminated.

The cyclic crack growth information presented in this paper was not calculated in accordance with E 647. In the case of the constant-ΔK data, a linear regression analysis of all the crack length information was used. For the constant cyclic load fatigue crack growth experiments, a procedure similar to the secant method was used. The secant method defines the crack growth rate as the slope of the line between two points on the crack length versus cycles curve spaced at the required Δa increment. The method used in this report was to fit a straight line to all the data obtained over the required Δa increment. For the data presented in Fig. 7, an increment of 0.5 mm was used. In this way, all the crack length information determined over the required increment is used in the rate determination. The procedures in E 647 should be modified in a way similar to this to take advantage of the large amount of information which can be obtained using automated procedures.

It must be understood, however, that the ability to acquire more data over a smaller interval of crack extension with the automated procedure does not imply that the automated experiment can be adequately performed over small crack extension intervals. In fatigue crack growth testing in general (whether automated or manually performed), a problem exists in that it is difficult to determine how far to propagate a crack in order to insure that the experiment has sampled a true cross section of the specimen and has not been localized to one (possibly unrepresentative) region of the material. The automated constant ΔK experiments described in this report were performed with 2 to 2.5

mm of crack extension since most transients have been observed to occur well within this interval. Therefore data within these intervals can be assumed to be representative of the bulk material behavior. Likewise, in the constant-load experiments described, the crack growth rates were calculated from data taken over an 0.5-mm increment in crack length (twice that required by E 647) for the same reasons. In general, the investigator must verify for each test that sufficient crack extension has occurred to truly sample the material.

Accuracy

It is difficult to determine the accuracy to which the cyclic crack growth rate has been determined. There are two major variables which contribute to possible errors in the measured crack growth rate: errors in the measured crack length, and errors in the range in stress-intensity factor. These two variables are not independent since the range in stress-intensity factor is a function of the crack length, the range in applied load, and the geometry of the test specimen. Since the accuracy of the crack length readings has the biggest impact on the accuracy of the cyclic crack growth rates, the accuracy of the crack length reading was of prime consideration in the development of the automated testing system described herein. In developing the specifications for the hardware, a mathematical model was developed. A description of the model and the results of the analysis can be found in Refs 1 and 2. Briefly, the model was used to study the effect of electrical noise and A/D resolution on expected crack length accuracies. An example of the type of results obtained from the model is presented in Fig. 9. The data shown in this figure are the expected 95% confidence interval on crack length, assuming that the Saxena and Hudak [3] compliance expression is exact for a 2T compact fracture specimen with ratios in a/w of 0.5 and 0.8. It was assumed that there was ±15 mV of electrical noise on the load signal and ±50 mV of electrical noise on the deflection signal. These noise levels are fairly typical of the levels present in the autoclave testing system. The calibration of the load signal is 10 V representing 8900 N while with the displacement signal 10 V represents 0.18 mm of deflection. As can be seen in this figure, the resolution in crack length under these conditions is a strong function of a/w and the ΔK level. Figures 10 and 11 present two examples of how well this model predicts the results of the experiments. In Fig. 10 is presented the results of an experiment conducted at a ΔK level of 33 MPa$\sqrt{m}$ and an a/w of 0.51. Under these conditions, very little scatter is observed in the test data similar, as predicted by the model. The model predicted a confidence interval less than ±0.064 for the individual data points, which was in fact observed.

In Fig. 11 the results of an experiment conducted at a ΔK level of 11 MPa$\sqrt{m}$ and an a/w of 0.50 are presented. Under these conditions, the scatter in the data is quite large, as was predicted (at least qualitatively) by the model. In this case, the model predicted a confidence interval of ±0.10 mm

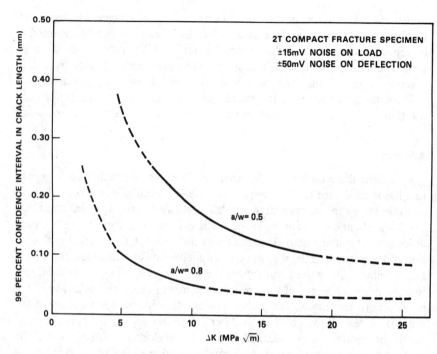

FIG. 9—*Results from crack length resolution model.*

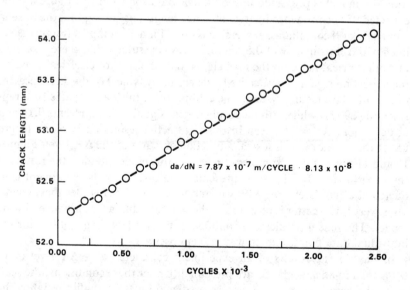

FIG. 10—*Crack length versus cycles data for fatigue crack growth experiment conducted at a constant-$\Delta K = 33$ MPa$\sqrt{m}$ on SA533B-1 material in 288°C PWR water environment.*

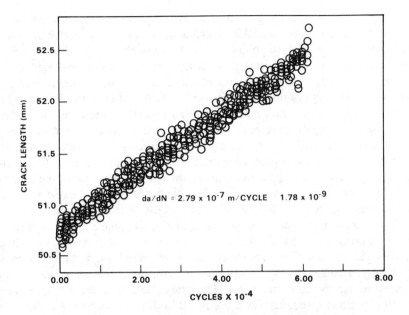

FIG. 11—*Crack length versus cycles data for fatigue crack growth experiment conducted at a constant-*Δ*K = 11 MPa$\sqrt{m}$ on SA533B-1 material in 288°C PWR water environment.*

while the scatter in the figure is approximately ±0.20 mm. Such a discrepancy between the predicted and observed scatter is not unreasonable since the model is intended to predict the degree of scatter that would be observed if the test were conducted under the best possible conditions. Any deviation from the optimum conditions (such as an increase in thermal noise within the test system) will cause a corresponding increase in scatter. In this vein, it should be understood that the data in Fig. 10 were obtained at a frequency of 0.1 Hz while the data in Fig. 11 were obtained at 0.01 Hz, suggesting that thermal fluctuations (or other relatively low-frequency noise components) not included in the model could in fact have had a greater influence on the data in Fig. 11 than those in Fig. 10.

It appears from this comparison that the model is a fair representation of what is occurring during the experiment. The results of this type of analysis are of little help in improving the data once the experiment has been conducted. However, such an analysis has proven to be very useful in developing specifications for the system hardware and in improving the possible resolution of a system.

The effect of this scatter in crack length measurements on the calculated crack growth rate and the consequences to the test procedure are clear from the results in Figs. 10 and 11. Note that while the data in Fig. 10 are clearly less scattered than those in Fig. 11, the 95% confidence intervals of the calculated slopes (that is, crack growth rates) are proportionally the same in both cases. This is due to the fact that the analysis in Fig. 11 was performed on a sufficient number of data points (many times more than that shown in Fig. 10) to average the scatter, as well as because the crack was propagated a distance much greater than the degree of scatter in the data. In the case of the automated testing system described here, therefore, if experiments are to be conducted in the ΔK range below 10 MPa$\sqrt{m}$, either the crack has to be propagated a long distance in order to get a reasonable estimate of the cyclic crack growth rate (which is extremely time-consuming and expensive at these low ΔK levels) or the resolution in crack length measurement has to be improved. From the model discussed above the only way to improve the resolution in the crack length measurement is to conduct the test at large a/w ratios or substantially reduce the electrical noise on the deflection signal.

The other source of error in the determination of cyclic crack growth rates involves errors in the range in stress-intensity factor (ΔK). As was discussed earlier, ΔK is a function of crack length, the range in load, and the specimen geometry. Besides errors in crack length, errors in specimen geometry can also cause significant errors in ΔK. The straightness of the crack front is part of the specimen geometry from the K calibration standpoint. Crack front straightness can be a significant problem especially when fatigue crack growth experiments are conducted in aggressive environments. The procedures described in E 647 for correcting for crack front curvature only correct the crack length determination; they do not correct for the effect of a curved crack front on the expression for the stress-intensity factor.

The error in cyclic crack growth rate due to crack front straightness can be significant because the stress-intensity factor at a point on the crack front is a function of the curvature, and the crack growth rate is proportional to the range in stress-intensity factor raised to a power close to four.

Conclusion

The following conclusions have been reached concerning the automation of the fatigue crack growth experiment.

1. The automated fatigue crack growth experiment allows personnel not skilled in computer programming to run this type of test.

2. Fatigue crack growth experimentation allows three different types of experiments to be conducted:

- constant-load tests with increasing stress-intensity factor,
- constant stress-intensity experiments, ΔK held constant, and
- decreased stress-intensity experiments, ΔK decreasing.

3. The automatic fatigue crack growth program automates the entire experiment from input of test condition through system checkout, actual test operation, and subsequent data analysis.

4. Test results indicate crack length measurements can be made with a relatively high degree of confidence.

5. Experiments in the ΔK range of <11 MPa$\sqrt{m}$ require a long crack propagation to achieve a reasonable estimate of cyclic crack growth rate.

6. The straightness of the crack front is an important factor in crack length measurement.

Acknowledgments

The authors wish to acknowledge the assistance of P. F. Harold of Babcock & Wilcox, who wrote most of the computer programs used in the system described in this paper, and R. L. Jones of the Electric Power Research Institute, whose support of the autoclave fatigue project made the work possible.

References

[1] Van Der Sluys, W. A., "Corrosion Fatigue Characterization of Reactor Pressure Vessel Steels—Phase I Report," Electric Power Research Institute Interim Report NP-2775, Palo Alto, CA, Dec. 1982.

[2] Van Der Sluys, W. A. and DeMiglio, D. S., *Sensitive Fracture: Evaluation and Comparison of Test Methods, ASTM STP 821*, American Society for Testing and Materials, Philadelphia, 1984, pp. 443–469.

[3] Saxena, A. and Hudak, S. J., Jr., "Review and Extension of Compliance Information for Common Crack Growth Specimens," Paper 77-9E7-AFCGR-P1, Westinghouse Scientific, Pittsburgh, PA, May 1977.

[4] Donald, J. K. and Schmidt, D. W., "Computer-Controlled Stress Intensity Gradient Technique for High Rate Fatigue Crack Growth Testing," *Journal of Testing and Evaluation*, Vol. 8, No. 1, Jan. 1980, pp. 19–24.

R. H. VanStone[1] and T. L. Richardson[1]

Potential-Drop Monitoring of Cracks in Surface-Flawed Specimens

REFERENCE: VanStone, R. H. and Richardson, T. L., "**Potential-Drop Monitoring of Cracks in Surface-Flawed Specimens,**" *Automated Test Methods for Fracture and Fatigue Crack Growth, ASTM STP 877,* W. H. Cullen, R. W. Landgraf, L. R. Kaisand, and J. H. Underwood, Eds., American Society for Testing and Materials, Philadelphia, 1985, pp. 148–166.

ABSTRACT: A direct-current potential-drop crack monitoring technique has been modified for use with test specimens having rectangular cross sections and semicircular crack starting notches, with radii as small as 0.075 mm (0.003 in.). An experimental program was conducted to evaluate an analytical model which predicts the crack depth from the potential drop, crack aspect ratio, and potential probe spacing. The fatigue crack growth rates of Rene´ 95, an advanced nickel-base superalloy, were determined at room temperature for three levels of gross section stress. There was extremely close agreement between the crack depths determined experimentally through post-test fractography and those predicted by the analytical potential-drop model. The critical barrier for fully automated monitoring of surface flaw crack growth with potential drop is the development of relationships to predict the crack aspect ratios as a function of material and testing parameters.

KEY WORDS: fatigue crack growth, test method, potential drop, surface flaws, superalloys

An important part of the material characterization and life management of many high-strength materials is the accurate determination of the fatigue crack growth rates as described using linear elastic fracture mechanics. In many service applications, it is critical to understand crack growth behavior for high-stress/short-crack conditions. One specimen commonly used to evaluate the fatigue crack growth rates in high-strength aircraft engine disk alloys [1–3] is known as the K_b specimen. This specimen geometry was originally developed by Johnson et al [1]. The K_b specimen has a rectangular-shaped gage section containing a semi-elliptical surface flaw, as shown schematically

---

[1]Engineer, Materials Life and Methods, and research and development specialist, respectively, General Electric Co., Aircraft Engine Business Group, Cincinnati, OH 45215.

in Fig. 1. In previous studies [1–3], the growth of surface cracks from semi-elliptical flaws was monitored using cathetometry. Tests conducted in a laboratory not staffed continuously often required test interruptions for periods up to two days, which occasionally resulted in small amounts of crack arrest.

This paper will describe the adaptation of a direct-current (d-c) potential-drop crack monitoring technique to this specimen geometry. The motivation for this study was to provide continuous monitoring of crack growth while increasing the quality and quantity of data, and reducing testing costs.

Background

The d-c potential drop-technique has been used for many years to monitor the crack growth rates in specimens with long cracks (that is, compact specimens) [4–6]. More recently, several investigators [7–9] have used the d-c potential-drop method to monitor cracks in specimens with short initial crack depths. Gangloff [7] initially developed a high-sensitivity potential-drop method to monitor the crack growth from 0.1016-mm-deep (0.004 in.) and 1.4224-mm-wide (0.056 in.) chord-shaped electric discharge machined (EDM) defects in axisymmetric, 5.08-mm-diameter (0.2 in.) fatigue specimens. Subsequently, VanStone et al [8] modified the Gangloff approach and monitored crack growth from 0.1016-mm-radius (0.004 in.) semicircular EDM notches in 6.35-mm-diameter (0.25 in.) fatigue specimens. Both of these investigations demonstrated that the Roe-Coffin [10] potential-drop solution predicted short crack lengths within 15%. The solution was derived from an analytical fluid flow solution [11] using a half-ellipsoid-shaped dis-

FIG. 1—*Schematic drawing of* K_b *specimen.*

continuity along a flat surface of an infinite fluid field. This situation more closely resembles the geometry of a surface crack in a K_b specimen as shown in Fig. 1 than the surface crack in a round bar considered by Gangloff [7] and VanStone et al [8]. Thus, no difficulties were anticipated in accurately predicting the crack depth using the Roe-Coffin solution.

Experimental Approach

The purpose of this investigation was to verify the applicability of a d-c potential-drop technique and the accuracy of the Roe-Coffin solution to predict crack depth in K_b specimens. In order to minimize variables which are known to alter crack growth rates, a single material was evaluated using the same temperature, stress ratio (R), and cyclic frequency. Restricting the evaluation to a single material and temperature does not imply that this technique is restricted to this set of experimental conditions. Previous work using this potential-drop technique on axisymmetric specimens [7,8] has shown that this technique can be successfully used on a variety of materials at temperatures from room temperature to 649°C (1200°F).

Material

The material used for this study was powder-metallurgy Rene´95 which was compacted to full density using hot isostatic pressing (HIP). This alloy is an advanced nickel-based superalloy which is used in aircraft engine disk applications. The specimens evaluated in this study were machined from a turbine compressor disk. A disk HIP container was loaded with −140-mesh powder and HIP-compacted at 1121°C (2050°F) and 103 MPa (15 ksi). It was then solution treated at 1127°C (2060°F), quenched into 538°C (1000°F) salt, and aged at 871°C (1600°F) for 1 h followed by a 649°C (1200°F) age for 24 h. Table 1 compares the composition of this disk with the Rene´95 specification. This composition is within the specification limits. The room temperature 0.2% offset yield strength and ultimate tensile strength of the material are 1213 and 1660 MPa (176 and 241 ksi), respectively.

Test Conditions

The K_b specimens used in this study had nominal gage widths and thicknesses of 15.24 and 6.35 mm (0.6 and 0.25 in.), respectively. Each specimen had a semicircular EDM notch introduced into the center of the large face of the gage section using a tantalum electrode. D-c potential probes were attached approximately 0.406 mm (0.016 in.) above and below the EDM notch. Figure 2 shows schematically the position of the EDM notch and potential probes. Also shown in this figure is the nomenclature used to describe the notch and potential probe spacing (L_p). The EDM notch depth (a_n) and half-

TABLE 1—*Composition of Rene '95 material.*

Element	Specification Limits, weight %	Check Analysis, weight %
Cr	12.0 to 14.0	13.11
Co	7.0 to 9.0	8.06
Mo	3.3 to 3.7	3.49
W	3.3 to 3.7	3.47
Al	3.3 to 3.7	3.46
Ti	2.3 to 2.7	2.42
Cb	3.3 to 3.7	3.46
C	0.04 to 0.09	0.06
Mn	0.15 max	0.015
Si	0.20 max	<0.01
Fe	0.50 max	0.15
P	0.015 max	0.003
S	0.015 max	0.004
B	0.006 to 0.015	0.014
N	0.005 max	0.002
H	0.001 max	0.0008
O	0.010 max	0.0066

surface length (c_n) are followed by the subscript n to indicate the initial notch dimensions. The same nomenclature was used to indicate crack depth (a) and half-surface length (c), except that the subscript was not used.

A total of five K_b specimen tests were conducted using conventional closed-loop electrohydraulic test equipment. The tests were run in a constant-load control mode. The maximum gross section stress for the five specimens varied from 689 to 1034 MPa (100 to 150 ksi). The R stress ratio (minimum load/maximum load) was zero and the specimens were cycled at 0.33 Hz (20 cpm) using a triangular wave shape. Each test was intentionally interrupted several times, and the specimen was heat tinted using induction heating. The purpose of this heat tint procedure was to mark the crack dimensions at several positions to allow post-test measurement of crack shape development.

During these tests, the electric potential was monitored using the potential-drop system shown schematically in Fig. 3. This system is identical to the one used by VanStone et al [8] and is similar to that used by Gangloff [7]. Direct current is passed through a specimen containing an EDM defect. Potential monitoring probes, located on either side of the defect, are used to measure the potential drop across the region containing the defect. The microcomputer is used to trip a transistor relay which applies voltage to the specimen. The current is turned off for one second during each fatigue cycle. This permits determination of the potential with and without applied voltage, and thus eliminates thermocouple effects. During each fatigue cycle, the microcomputer monitors these two values of potential and subtracts them to determine the potential difference (V). The maximum and minimum loads are also monitored for each fatigue cycle. At selected intervals the potential and load values

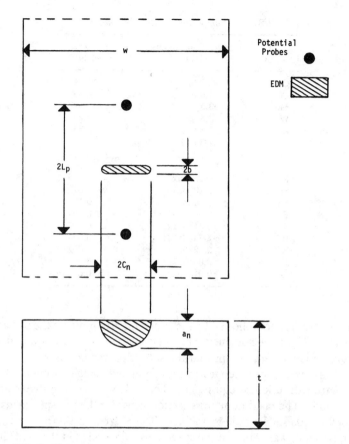

FIG. 2—*Nomenclature of specimen notch, and potential lead dimensions.*

are averaged, printed at a teletype, and recorded on a magnetic tape for future analysis using a conventional time-sharing computer system.

A current of 30 A was used on these specimens, resulting in an initial potential difference at the EDM notch (V_n) of approximately 350 μV. The potential increases as the crack emanating from the notch increases in area. The electrical potential difference (V) is normalized by the microcomputer to the initial potential difference at the notch (V_n) to eliminate the effects of alloy resistivity, temperature, specimen cross-sectional area, applied current, and specimen insulation variables. The V/V_n values were also recorded on the magnetic tape for data analysis.

Data Analysis

In each of the tests performed in this investigation, approximately 1000 data sets were acquired on magnetic tape. These data included cycle number,

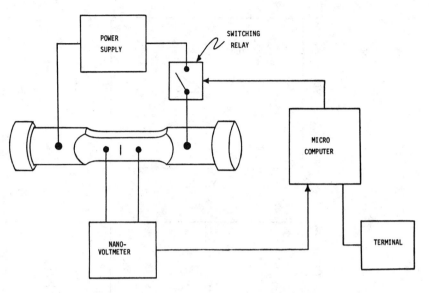

FIG. 3—*Schematic diagram of d-c potential-drop system.*

loads, V, and V/V_n. Comparison of the potential for the data set immediately preceding a heat tint with that for the subsequent data set often showed a small but significant increase in V/V_n as shown schematically in Fig. 4a. The change in V/V_n was typically less than 0.5%; however, multiple heat tints can result in changes of several percent during the test. This shift is probably caused by annealing of the potential lead probes and wires and by oxidation of the crack faces. To avoid the type of discontinuity shown in Fig. 4a, the V_n value used for the segment of data following the heat tint was modified so that the V/V_n values from the blocks preceding and succeeding the heat tint cycle were identical. This resulted in a V/V_n-cycle relationship as shown schematically in Fig. 4b.

Based on experience of potential-drop tests with axisymmetric specimens [7,8], it was desirable to reduce the total number of data points to approximately 100. Previously [7,8], the potential and load values for a constant number of data sets were averaged; for example, a total of 1000 data sets were reduced to 100 sets by block averaging each increment of 10 data sets. As is typical for load-controlled crack propagation tests, the crack accelerates during the test. Averaging the test data on the basis of cycles results in increasing the difference in potential and hence crack depth between each block averaged data set.

A more useful block averaging approach would be to average the load and potential data on the basis of increments in crack depth. This would result in decreasing block size with increasing crack depth. This procedure would also extend the range of stress intensity (ΔK) over which crack growth rate

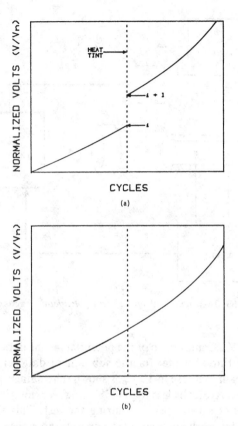

FIG. 4—*Schematic diagram showing* (a) *potential data and* (b) *adjusted potential data to eliminate heat tint transients.*

(da/dN) data would be determined when the seven-point sliding polynomial technique [ASTM Test for Constant-Load-Amplitude Fatigue Crack Growth Rates Above 10^{-8} m/Cycle (E 647-81)] is used. The sliding polynomial approach does not report da/dN or ΔK data for the first three or last three data points. If crack depth is not sampled frequently when a crack is growing rapidly, exclusion of the last three data points may significantly reduce the range of available da/dN and ΔK data. For a semicircular-shaped crack, the normalized potential increases with an almost linear relationship with crack depth [8], particularly at larger values of crack depth. Thus, for the case of block averaging data, crack depth can be considered to vary linearly with normalized potential (V/V_n). The variable block average procedure developed during this investigation is shown schematically in Fig. 5. The range of V/V_n was divided into n equal increments. In the first increment of data sets, n sets of load and potential data were averaged. In each of the subsequent intervals of V/V_n, the number of data sets averaged was reduced by one, so that for the

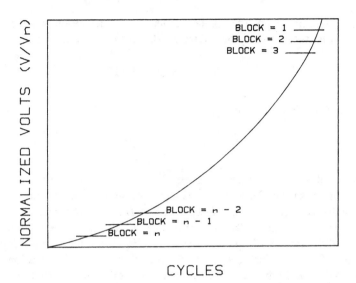

FIG. 5—*Schematic diagram showing the variable block averaging procedure.*

final increment of V/V_n, the individual data sets were not averaged. The value of n was selected to result in approximately 100 datum points per test. These data were renormalized to V/V_n were the minimum potential difference was considered to be V_n [8].

The next step in the data analysis is to convert the V/V_n potential data to crack length data using the Roe-Coffin solution. The total solution is described in Appendix I. Although this solution is complex and includes elliptic integrals, it can be summarized by

$$\frac{V}{V_n} = f(a, c, a_n, c_n, b_n, L_p) \tag{1}$$

where f is the function of the appropriate notch dimensions (a_n, c_n, b_n), the crack dimensions (a, c), and the potential probe spacing (L_p). Fixing V/V_n, L_p, and the notch dimensions leaves both a and c as independent variables. In both the potential solution and K-solution, the crack aspect ratio (c/a) is an independent variable. The previous studies [7,8] on monitoring the growth of short cracks assumed empirical relationships between c and a. In this investigation, the heat tinting procedure previously described was used to monitor the crack shape. Based on the work of Adams [12], for K_b specimens monitored by cathetometers, it was assumed that

$$c = A_1 + A_2a + A_3a^2 \tag{2}$$

where A_1, A_2, and A_3 are adjustable coefficients. By definition, when $c = c_n$, $a = a_n$. Substitution of this boundary condition into Eq 2 results in

$$A_1 = c_n - A_2 a_n - A_3 a_n^2 \tag{3}$$

Substitution of Eq 3 into Eq 2 yields

$$c = c_n + A_2(a - a_n) + A_3(a^2 - a_n^2) \tag{4}$$

The values of A_2 and A_3 were determined by performing a least-squares regression analysis to the form of Eq 4 using the values of a_n, c_n, a, and c experimentally determined from post-test fracture surface measurements.

Using the experimentally determined V/V_n data, regression analysis (Eq 4), and the Roe-Coffin potential solution (Eq 1), a value of crack depth is determined for each block averaged data set using an iterative solution technique. The crack depth and cycle number data are then converted to da/dN data using the seven-point sliding polynomial technique (ASTM E 647-81).

The K-solution used during this investigation was that originally developed for the K_b specimen by Johnson et al [1]. For completeness, this K-solution is described in Appendix II. The appropriate K-values for the specimens evaluated in the present study were determined from the crack depths as defined in the seven-point sliding polynomial approach (E 647-81), the surface crack lengths as defined in Eq 4, and the K-solution in Appendix II.

Results

Table 2 lists the applied stress, EDM notch dimensions, final crack dimensions, and a summary of the heat tint data for the five Rene´ 95 specimens tested in this study. The correlation coefficients listed in Table 2 for the aspect ratio relationship of Eq 2 are extremely high, indicating excellent agreement between the predicted and experimentally measured crack surface lengths. As an example, Fig. 6 shows the variation of c with a and the result of the regression analysis for Specimen 4.

The final aspect ratio (c/a) data in Table 2 show that as the applied stress is increased, the crack tunnels into the specimen, resulting in diminished c/a values. Specimens 2, 4, and 5 were all cycled with a maximum stress of 689 MPa (100 ksi) and had final crack aspect ratios (c/a) ranging from 0.96 to 1.02. Specimen 3 with a maximum stress of 827 MPa (120 ksi) had a slightly lower aspect ratio. Specimen A5, the most highly stressed specimen, had an aspect ratio of 0.76. The variation of c/a with crack depth is shown in Fig. 7 for all five specimens. The lines in Fig. 7 were constructed by substitution of the regression coefficients A_1, A_2, and A_3 into the algebraic manipulation of Eq 2

$$c/a = \frac{A_1}{a} + A_2 + A_3 a \tag{5}$$

This figure illustrates the strong influence of applied stress on aspect ratio.

Using these aspect ratio relationships, the Roe-Coffin analytical potential solution was used to predict the crack depths. Figure 8 shows the crack depths predicted from the potential solution as a function of the crack depths determined using post-test measurements of heat tint markings. The line in this figure has a slope of one. If the predictions were perfect, all the data points would fall on this line. It should be noted that the potential solution predicts the crack depth from V/V_n, the notch dimensions, and the aspect ratio relationship with no other adjustable parameters. Statistical analysis of these data shows that the average error is -0.0203 mm (-0.0008 in.) with a standard deviation of 0.0406 mm (0.0016 in.). This extremely close agreement shows that the Roe-Coffin model accurately predicts crack depths from the potential data when the crack aspect ratio is known.

The importance of the aspect ratio can be observed by plotting the analytically-determined crack depth as a function of normalized potential as shown in Fig. 9. Specimens tested with a given stress had extremely similar relationships between potential and crack depth. This results primarily from the change in aspect ratio with applied stress.

The fatigue crack growth rate data for all five specimens are shown in Fig. 10. No data are shown from crack depths within 0.0508 mm (0.002 in.) of the EDM notch because those data may be influenced by the notch geometry. The ΔK values were calculated using the K-solution given in Appendix II and a cyclic yield stress of 1469 MPa (213 ksi). This value was estimated from

TABLE 2—Summary of Rene '95 K_b specimen tests.

	Specimen No.				
	2	3	4	5	A5
Maximum stress, ksi	100.0	120.0	100.0	100.0	150.0
a_n, in.	0.0042	0.0067	0.0041	0.0041	0.0090
c_n, in.	0.0040	0.0066	0.0040	0.0041	0.0088
L_p, in.	0.0160	0.0160	0.0160	0.0160	0.0160
a_f, in.	0.0375	0.0515	0.0747	0.0638	0.0930
c_f, in.	0.0383	0.0473	0.0725	0.0615	0.0705
c_f/a_f	1.02	0.92	0.97	0.96	0.76
No. of heat tints	3	3	3	4	2
A_1, in.	-0.001006	-0.000695	-0.000467	-0.000370	-0.000711
A_2, in./in.	1.210076	1.112276	1.096120	1.099421	1.087968
A_3, 1/in.	-4.319042	-3.508373	-1.589179	-2.220452	-3.465136
Correlation coefficient	1.00000	0.99940	0.99997	0.99461	1.00000

Conversion factors: 1 in. = 25.4 mm; 1 ksi = 6.894757 MPa.

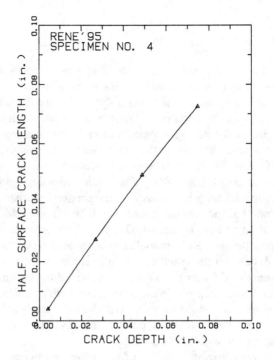

FIG. 6—*Variation of half-surface length* (c) *with crack depth* (a) *for specimen 4.*

the monotonic yield strength. There is very close agreement between these data even though there were substantial variations in applied stress and aspect ratios. This close agreement results primarily from the ability to predict crack depth using potential drop as shown in Fig. 8. The use of a different K-solution may cause a relative shift in the crack growth data [13], but will not cause a major difference between the tests.

Discussion

These results have shown that the use of this potential-drop technique, combined with the Roe-Coffin potential solution, accurately predicts crack depths. The chief limitation of this potential-drop technique is the requirement to determine crack shape changes with crack depth. This study used regression analyses of crack shapes determined from heat tints. Although this procedure worked well, it reduced testing efficiency and may, in certain cases, cause transients in crack growth behavior. An alternative approach would be to develop empirical crack aspect ratio relationships. This process has been used previously on other specimen geometry or EDM notch configurations [7,8,14].

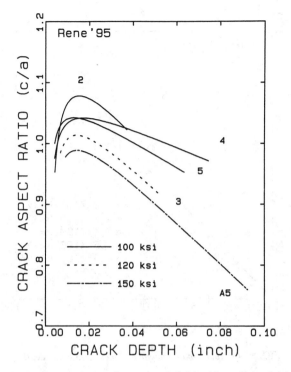

FIG. 7—*Variation of crack aspect ratio* (c/a) *with crack depth* (a).

This study showed that crack shape is sensitive to the applied stress. These observations can be rationalized based on recent analytical work of Trantina, deLorenzi, and Wilkening [15,16]. They performed three-dimensional elastic-plastic finite-element analyses of semicircular and semi-elliptical (c/a = 1.4) surface flaws in a semi-infinite medium under the conditions of monotonic loading. Under totally elastic conditions, where the maximum stress is less than 0.7 of the yield strength, K at the specimen surface exceeds that in the interior. At higher stress levels, the lack of constraint at the specimen surface results in a larger plastic zone size and lower K at the surface, as calculated from the J-integral, than that at the crack front on the crack symmetry line. The lower K-value at the specimen surface relative to the interior position results in lower crack growth rates at this location. Thus, the c/a ratio is reduced as crack depth is increased. This is shown by the overall trend in Fig. 7 for the Rene´95 evaluated in this program. Trantina and co-workers [15,16] also showed that the stress intensity at the surface relative to that along the crack symmetry line diminished with increasing applied strain. Their results can be used to rationalize why the cracks tend to have reduced aspect ratios with increasing applied stresses.

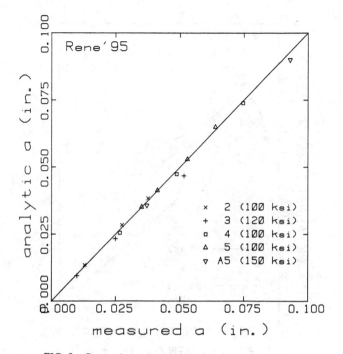

FIG. 8—*Comparison of analytic and measured crack depths.*

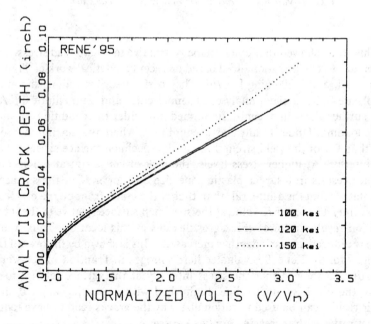

FIG. 9—*Variation of analytical crack depth (a) with normalized potential (V/V$_n$).*

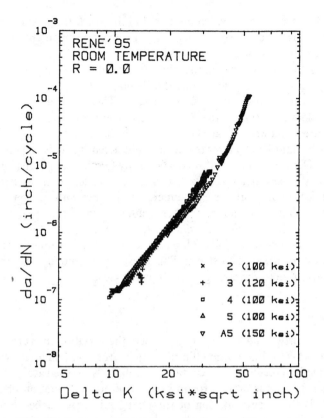

FIG. 10—*Fatigue crack growth rate data for Rene '95 at room temperature.*

The finite-element analysis [15,16] was based on monotonic loading while the observations of the Rene´ 95 study were determined from cyclically loaded specimens. This difference may alter the magnitude of the diminished K at the specimen surface, but the diminished K-values due to loss of constraint from yielding are still likely to occur under fatigue loading.

It may be possible to perform a finite-element analysis similar to that performed by Trantina et al [15,16] to predict the changes in crack shape during a test; however, this approach would require both monotonic and cyclic constitutive data for an accurate analysis.

At the present time, neither an empirical aspect ratio relationship nor finite-element solutions have been developed to model the variations in crack shape. This is especially true for evaluation of crack growth rates of new materials or new testing conditions of more well-characterized materials. An example of this type of test is the monitoring of crack growth under hold-time conditions where cracks tend to tunnel more than in continuously cycled tests [17]. This presumably results from reduced constraint at the specimen surface resulting from creep deformation within the crack-tip plastic zone. Due to these concerns, near-term testing of K_b specimens with potential drop will use heat tinting to define the aspect ratio.

This investigation has shown that the potential-drop technique can be used to rapidly and accurately determine the Region II or Paris Law portion of a fatigue crack growth curve in K_b specimens. This test method is not well suited for determination of the Region I crack growth rates for several reasons. Figure 9 shows that at short crack depths, small increases in crack depth do not significantly alter the normalized potential. This response is typical for cracks with aspect ratios close to unity [8]. A much higher potential sensitivity is observed for cracks with high c/a ratios, such as the chord-shaped defect in axisymmetric specimens [7,8]. Even with the low potential sensitivity at short crack lengths, other techniques to determine near-threshold crack growth, such as K-shedding [9], could be used. This approach is complicated in specimens containing semi-elliptical cracks due to the change in aspect ratio with applied stress. This is especially true for near-threshold crack growth due to the strong effect of crack closure.

Conclusions

The Roe-Coffin model accurately predicts the variation in crack depth in K_b specimens containing semi-elliptical cracks from d-c potential-drop data. This test method results in an accurate and relatively inexpensive way to monitor Region II fatigue crack growth rates. The major limitation of this test technique is the documentation of the changes in crack aspect ratio with crack size. This crack tunneling behavior most likely results from retarded crack growth rates at the specimen surface. At this location, the lack of constraint permits a larger plastic zone size and thus a reduced value of K.

Acknowledgments

The authors appreciate the support and encouragement of D. A. Utah and L. L. Walker throughout this study. P. K. Wright graciously provided the specimens for this investigation. Discussions with L. T. Duvelius and J. H. Adams were extremely helpful. The authors greatly appreciate the assistance of D. D. Krueger in reviewing this manuscript.

APPENDIX I

This Appendix describes the Roe-Coffin potential solution for a semi-elliptical flaw having the geometry shown in Fig. 2. This solution was used to determine the crack geometries from the normalized potential. The potential (V)

$$V = V_0 L_p \frac{[\sqrt{1 - k^2 \sin^2 \theta}/(\tan \theta)] + E(k, \theta) - Q}{E(k, \pi/2) - Q}$$

where

$$V_0 = \text{remote potential,}$$

$$E(\delta_1, \delta_2) = \int_0^{\delta_2} (1 - \delta_1^2 \sin^2 \phi)^{1/2} \, d\phi,$$

$$\theta = \tan^{-1}(\sqrt{a}),$$

$$\theta_0 = \tan^{-1}(b/\beta), \text{ and}$$

$$Q = E(k, \theta_0) + \frac{b\beta^2}{ac\gamma}.$$

There are two solutions for α, β, γ, and k depending on whether the crack aspect ratio is greater or less than unity. Within those two solutions the value of α varies depending on the relative difference between crack dimensions and probe spacing. The parameter z is the dimension by which the probe spacing is off the centerline of EDM defect. As in previous studies [7,8,13], it was assumed that z equals 0.0127 mm (0.0005 in.).

For $c \geq a \geq b$

$$\beta^2 = a^2 - b^2$$

$$\gamma^2 = c^2 - b^2$$

$$k^2 = 1 - \beta^2/\gamma^2$$

If $L_p^2 + z^2 \geq \gamma^2$

$$\alpha = \frac{1}{2} \left[\frac{L_p^2 + z^2 - \gamma^2}{\beta^2} + \sqrt{\left(\frac{L_p^2 + z^2 - \gamma^2}{\beta^2} \right)^2 + \frac{4\gamma^2 L_p^2}{\beta^4}} \right]$$

If $L_p^2 + z^2 < \gamma^2$

$$\alpha = \frac{2\gamma^2 L_p^2/\beta^4}{\dfrac{\gamma^2 - L_p^2 - z^2}{\beta^2} + \sqrt{\left(\dfrac{\gamma^2 - L_p^2 - z^2}{\beta^2} \right)^2 + \dfrac{4\gamma^2 L_p^2}{\beta^4}}}$$

For $a > c > b$

$$\beta^2 = c^2 - a^2$$

$$\gamma^2 = a^2 - b^2$$

$$k^2 = 1 - \beta^2/\gamma^2$$

If $L_p^2 + z^2 \geq \beta^2$

$$\alpha = \frac{1}{2} \left[\frac{L_p^2 + z^2}{\beta^2} - 1 + \sqrt{\left(\frac{L_p^2 + z^2}{\beta^2} - 1 \right)^2 + \frac{4L_p^2}{\beta^2}} \right]$$

If $L_p^2 + z^2 < \beta^2$

$$\alpha = \frac{2L_p^2/\beta^2}{1 - \dfrac{L_p^2 + z^2}{\beta^2} + \sqrt{\left(1 + \dfrac{L_p^2 + z^2}{\beta^2}\right)^2 + \dfrac{4L_p^2}{\beta^2}}}$$

The only unmeasured parameter in this solution is V_0; however, the potentials are normalized by V/V_n so V_0 can be eliminated.

Close to the EDM notch, the influence of the height (b) of the EDM notch was considered. The algorithm used is that described by Gangloff [7] where the total potential V is a function of the potential solution for a crack (V_{ca}), the potential of a notch (V_n), and the potential of a crack having the size of the notch (V_{cn}):

$$V = V_{ca} + Q(V_n - V_{cn})$$

For $a_n > a > 2a_n$

$$Q = \left(2 - \frac{a}{a_n}\right)$$

For $a > 2a_n$

$$Q = 0$$

The value of b used to calculate V_{ca} and V_{cn} was zero. The experimentally measured b was used to calculate V_n.

APPENDIX II

K-Solution for K_b Specimen

The K-solution used to calculate the values of K shown in Fig. 10 is primarily that of Johnson et al [1] for the case of the maximum crack depth (a) within the specimen. The nomenclature used here is identical to that shown in Fig. 2

$$K = \frac{\sigma\sqrt{\pi a}}{\Phi} F_1 F_2 F_3 F_4$$

where

Φ = elliptic integral of second kind,
F_1 = front surface correction,
F_2 = loss of load-bearing area correction,
F_3 = plastic zone correction, and
F_4 = back surface correction.

The equations for these corrections are

$$F_1 = 1.1 - 0.07\left(\frac{a}{c}\right)$$

$$F_2 = \frac{1}{4}\left(3 + \frac{A}{A - A_c}\right)$$

$$F_3 = \left(1 - \frac{\sigma}{\sigma_y}\right)^{-0.05}$$

$$F_4 = 1.0 + \left(\frac{a}{t}\right)^{2.5} \exp\left[-2.5\,\frac{a}{c}\right]$$

where

t = specimen thickness,
w = specimen width,
A = specimen area = tw,
A_c = crack area = $\pi ac/2$, and
σ_y = cyclic yield stress.

The relationship for F_2 is slightly different than that used by Johnson et al [1]. The F_2 relationship just given uses the expression proposed by Williams and Isherwood [18]. The approximations used to estimate are

For $a/c > 1.0$

$$\Phi = \sqrt{\frac{a}{c}}\left[1.0 + 0.5708\left(\frac{a}{c}\right)^{-1.42}\right]$$

For $a/c \le 1.0$

$$\Phi = \left[1.0 + 0.5708\left(\frac{a}{c}\right)^{1.42}\right]$$

References

[1] Johnson, R. E., Coles, A., and Popp, H. G., *Journal of Engineering Materials and Technology*, Vol. 98, 1976, p. 305.
[2] Shahani, V. and Popp, H. G., "Evaluation of Cyclic Behavior of Aircraft Turbine Disk Alloys," NASA-CR-159433, National Aeronautics and Space Administration, Washington, DC, June 1978.
[3] Cowles, B. A., Warren, J. R., and Haake, F. K., "Evaluation of the Cyclic Behavior of Aircraft Turbine Disk Alloys, Part II," NASA-CF-165123, National Aeronautics and Space Administration, Washington, DC, July 1980.
[4] Wei, R. P. and Brazill, R. L. in *Fatigue Crack Growth Measurement and Data Analysis, ASTM STP 738*, American Society for Testing and Materials, Philadelphia, 1981, pp. 103–119.
[5] Ritchie, R. O. and Bathe, K. J., *International Journal of Fracture*, Vol. 15, 1979, pp. 47–55.
[6] Johnson, H. H., *Materials Research and Standards*, Vol. 5, 1965, pp. 442–445.
[7] Gangloff, R. P., *Fatigue of Engineering Materials and Structures*, Vol. 4, 1981, p. 15.
[8] VanStone, R. H., Krueger, D. D., and Duvelius, L. T., Fracture Mechanics, 14th Symposium, Vol. II: Testing and Application, *ASTM STP 791*, American Society for Testing and Materials, Philadelphia, 1983, pp. II-553–II-578.
[9] Wilcox, J. R. and Henry, M. F., Unpublished Research, General Electric Corporate Research and Development, Schenectady, NY.

[10] Roe, G. M. and Coffin, L. F., Unpublished Research, General Electric Corporate Research and Development, Schenectady, NY.

[11] Milne-Thomson, L. M., *Theoretical Hydrodynamics*, MacMillan, New York, 1963, pp. 505-511.

[12] Adams, J. H., Unpublished Research, General Electric Aircraft Engine Business Group, Cincinnati, OH.

[13] Gangloff, R. P. in *Fatigue Crack Growth Measurement and Data Analysis, ASTM STP 738*, American Society for Testing and Materials, Philadelphia, 1981, pp. 120-138.

[14] Gangloff, R. P., *Advances in Crack Length Measurement*, Engineering Materials Advisory Services, Ltd., London, 1982, pp. 175-230.

[15] Trantina, G. G. and deLorenzi, H. G. in *Proceedings*, Army Symposium on Solid Mechanics—Critical Problems in System Design, MMRC MS 82-4. Sept. 1982, pp. 203-214.

[16] Trantina, G. G., deLorenzi, H. G., and Wilkening, W. W., *Engineering Fracture Mechanics*, accepted for publication.

[17] Krueger, D. D., Unpublished Research, General Electric Aircraft Engine Business Group, Cincinnati, OH.

[18] Williams, J. G. and Isherwood, D. P., *Journal of Strain Analysis*, Vol. 3, 1968, pp. 17-22.

John J. McGowan[1] and J. L. Keating[1]

A Microprocessor-Based System for Determining Near-Threshold Fatigue Crack Growth Rates

REFERENCE: McGowan, J. J. and Keating, J. L., **"A Microprocessor-Based System for Determining Near-Threshold Fatigue Crack Growth Rates,"** *Automated Test Methods for Fracture and Fatigue Crack Growth, ASTM STP 877*, W. H. Cullen, R. W. Landgraf, L. R. Kaisand, and J. H. Underwood, Eds., American Society for Testing and Materials, Philadelphia, 1985, pp. 167–176.

ABSTRACT: A microcomputer-based system has been developed to perform near-threshold fatigue crack growth testing. The crack length is measured by both potential-drop and compliance techniques. The potential-drop technique has a resolution of 0.0005 W and is used for real-time load control. The compliance technique has a resolution of 0.002 W and is used for comparative reference. The potential-drop system uses d-c current and has multiple voltage probes (located remote from current input points) to minimize the effects of current variation and resistance changes. The potential-drop measurements are made with and without excitation to minimize thermal offset problems. Compliance and potential-drop measurements are determined at specified load points during a cycle to minimize the effects of crack closure.

The system is programmed to control the rate of ΔK change (for increasing or decreasing ΔK tests) to a user-selected value. As the applied load is changed, the sample points (for measuring compliance and potential drop) during the cycle are changed automatically. A real-time sliding-seven-point fatigue crack growth rate calculation is performed for each sample. The sampling frequency is adjusted by the system during the test to maintain measurement accuracy. The microcomputer records crack lengths, loads, cycle number, and specimen temperature on floppy disks for post-processing.

Near-threshold fatigue crack growth results are presented for Type 304 stainless steel at both 24 and 538°C.

KEY WORDS: fatigue crack growth, microcomputer, fracture mechanics, threshold

Nomenclature

a Crack length

B Specimen thickness

[1]The University of Alabama, College of Engineering, Tuscaloosa, AL 35486. Author McGowan is now a research engineer with the Oak Ridge National Laboratory, Oak Ridge, TN 37831.

da/dN Crack growth rate

E Young's modulus

e_0 Potential from lead at $x/W = 0.0$

$e_{0.25}$ Potential from lead at $x/W = 0.25$

ΔK Range of stress-intensity factor

P_{min} Minimum load

P_{max} Maximum load

R Load ratio, P_{min}/P_{max}

T_e Potential trigger load

T_{Lo} Low-compliance trigger load

T_{Hi} High-compliance trigger load

W Specimen width

x Distance measured from load line on specimen

All structures contain some imperfections or cracks, although flaws in modern structures are very limited in size due to advanced melting, welding, and inspection practices. The concepts of fracture mechanics are used routinely to guarantee that these small flaws do not grow to a dangerous level during operation of the component. The rate of crack growth per cycle (da/dN) is correlated by the range of the stress-intensity factor (ΔK). The largest fraction of the life of the component takes place with low crack growth rates ($da/dN = 10^{-8}$ m/cycle) with corresponding low ΔK levels. Therefore, it is this near-threshold crack growth region which has the most importance to the designer.

This region of crack growth is also the most difficult to obtain, as described by Bucci [1]. The primary reasons for this are (1) the need to reduce loads as the crack grows, and (2) the long duration of the tests. The load must be reduced at a controlled rate so as to eliminate transient conditions and thus erroneous results. Automatic, continuous load reduction using digital computer control yields accurate results in an efficient manner as described by Saxena et al [2]. This system requires little attention and can operate unattended for long periods.

This paper describes a relatively inexpensive microcomputer system which continuously reduces the load and performs real-time data analysis. The system utilizes d-c potential and compliance techniques for crack length measurement. Pertinent data are recorded on floppy disks for post-processing and plotting.

Remote Crack Length Measurement

One of the most important aspects of the microcomputer testing system was the remote measurement of crack length. The measurement techniques were required to be accurate, stable, and compatible with the microcomputer. Based upon a previous study conducted by Carden [3], two such measurement techniques were selected: d-c potential and crack-opening displacement (COD). Both systems have been extensively studied for remote crack length measurement. These two systems are sensitive to the average crack length and are generally unaffected by crack tunneling.

Direct Current Potential System

A fairly comprehensive review and assessment of potential systems has been compiled by Wei and Brazill [4]. In this review both a-c and d-c systems were studied. Either system was found to be satisfactory, each with its own peculiarities. Because of the availability of components and general simplicity, the d-c system was selected for use with the microcomputer. The d-c system described by Wei and Brazill used a fixed reference signal and one set of potential sense leads located on the front face of the specimen. This system required a precision, highly stabilized power supply and assumes a constant specimen resistance. For high-temperature testing, the resistivity can vary widely; therefore, Carden [3] advocated use of a signal from a reference specimen with the same temperatures and current as the test specimen. Ratioing the signal from the reference and test specimens gave a result that was independent of input current and temperature variations. Carden [3] also investigated the placement of the potential leads at points along the line of symmetry as shown in Fig. 1. In the present study the authors used two sets of leads on the test specimen to give an active potential reference. The two sets of leads are located as shown in Fig. 2, one at $x/W = 0\%$ and the other at $x/W = 25\%$. Using data from Fig. 1 the crack length is determined from the ratio $e_0/(e_0 - e_{0.25})$ as shown in Fig. 3. Note that the correlation is nearly linear throughout the entire data range and is usable up to $a/W = 85\%$.

A d-c system is generally sensitive to thermal electromotive force (emf) and other zero shifts. However, this problem is avoided by subtracting the potential reading at zero current. (Zero current readings are performed 20 s before each active reading.) As mentioned by Wei and Brazill [4], crack closure during fatigue can yield erroneous estimates of crack length as shown in Fig. 4. This error is avoided by measuring the potentials at some trigger load (T_e) when the crack is open. The resulting resolution of the d-c potential system is 5×10^{-4} W. (The resolution was determined by several hundred replicate readings at the same crack length.)

Crack-Opening Displacement System

Methods for using COD to determine crack length are well documented [5,6] and are used routinely in R-curve development [ASTM Standard Test for J_{Ic}, A Measure of Fracture Toughness (E 813-81)]. Due to the potential instrumentation and high temperature application, the COD measurement points were located at 0.625 W in from of the load line as shown in Fig. 2.

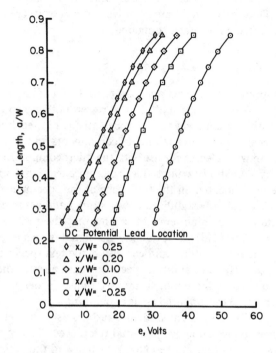

FIG. 1—*Variation of d-c potential with crack length and probe location.*

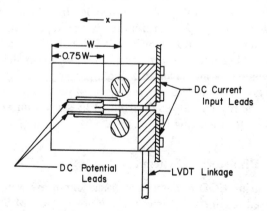

FIG. 2—*Test specimen geometry with LVDT linkage and potential probes indicated.*

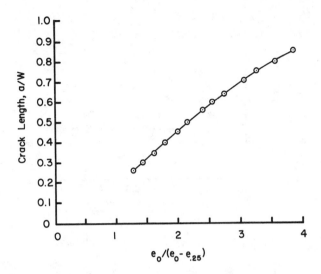

FIG. 3—*Direct-current potential versus crack length calibration.*

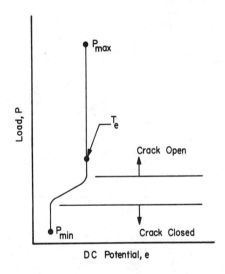

FIG. 4—*Schematic of load versus d-c potential trace with trigger level indicated.*

(The COD is measured by a linear variable differential transducer (LVDT) attached to rods away from the specimen). The calibration of $EB(COD)/P$ versus a/W was developed from the Ref 6 and is shown in Fig. 5.

As discussed by Yoder et al [7], the load displacement trace is generally nonlinear due to crack closure as shown in Fig. 6. To minimize errors due to

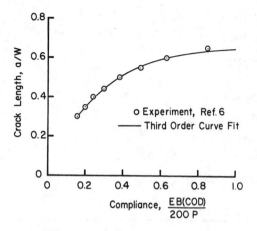

FIG. 5—*Compliance versus crack length calibration.*

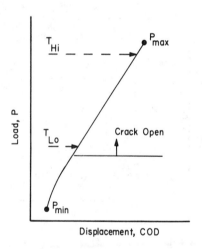

FIG. 6—*Schematic of load versus crack-opening-displacement trace with trigger levels indicated.*

crack closure, the compliance [(COD)/P] is determined by sampling the COD and P at two trigger loads (T_{Hi} and T_{Lo}) above the crack-opening load. Since the COD versus P is linear above the closure load, then the compliance is simply COD/P = (COD$_{Hi}$ − COD$_{Lo}$)/(P_{Hi} − P_{Lo}). The trigger levels are controlled by the microcomputer and the compliance is determined automatically by an MTS Model 438 Crack Correlator.

Microcomputer Control System

The control system utilizes an Apple II+ 64K system with Tecmar 12 bit analog/digital (A/D) and digital/analog (D/A) peripherals. The schematic of the system is shown in Fig. 7. The d-c potential crack length measurement system is composed of the low-voltage power supply and the high-gain amplifier. The D/A converter is used to turn the power supply on and off so that thermal emf can be offset. The MTS Crack Correlator is used to determine the specimen compliance. The COD and P signals are received from the MTS 442 Controller, and the load trigger levels come from the D/A converter. An MTS 442 Controller is used to regulate the specimen load and to provide full-scale control signals. The magnitude of the steady-state load (P_{mean}) and alternating load (ΔP) is controlled by the D/A converter. The Tecmar A/D converter takes 0 to 10 V analog signals from the high-gain amplifier, temperature indicator, cycle counter, and MTS Crack Correlator, and feeds

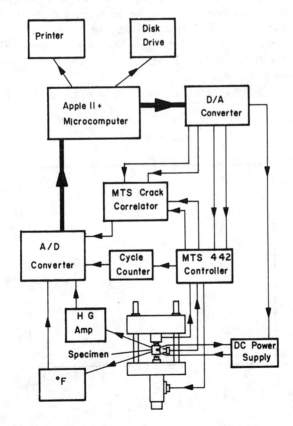

FIG. 7—*Schematic of microcomputer control system.*

the corresponding digital signal to the Apple II+. The A/D converter is synchronized to measure the d-c potential after the crack is open as shown in Fig. 4. The Tecmar D/A receives digital signals from the Apple II+ and passes the corresponding analog voltage to the MTS 442 Controller for ΔP and P_{mean}, and to the MTS Crack Correlator for T_{Hi} and T_{Lo}. The Apple II+ microcomputer receives information from the Tecmar A/D and issues commands through the Tecmar D/A.

Initial input information from the keyboard includes specimen dimensions, initial ΔK level, load ratio, and trigger load levels (T_{Hi} and T_{Lo}), final ΔK level, and ΔK decay rate. The microcomputer determines the necessary load levels and calculates the crack growth rates. Sampling frequency is determined by the actual crack growth rate so that the required measurement accuracy [ASTM Standard Test Method for Constant-Load-Amplitude Fatigue Crack Growth Rates Above 10^{-8} m/Cycle (E 647-81)] is maintained. The loads are determined continuously to control ΔK and the ΔK decay rate desired. A seven-point "incremental polynomial" curve fit [7] is used to determine the fatigue crack growth rate as the test is performed. The load trigger levels for compliance are fixed at a percentage of P_{max} and are adjusted automatically as the maximum load changes. The microcomputer prints and records on floppy disks the crack lengths, loads, cycle number, and temperature as the test proceeds.

The microcomputer system is well adapted for near-threshold fatigue crack

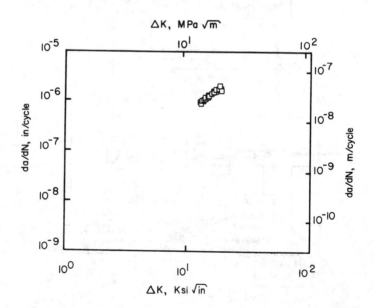

FIG. 8—*Direct-current potential-based fatigue crack growth rate results for Type 304 stainless steel at* R = 0.1 *and* 24°C.

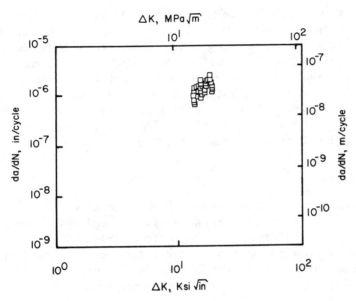

FIG. 9—*Compliance-based fatigue crack growth rate results for Type 304 stainless steel at* R *= 0.1 and 24°C.*

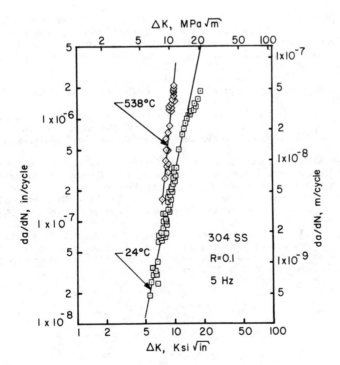

FIG. 10—*Fatigue crack growth rate data produced by microcomputer control system.*

growth measurement because it can decrease ΔK in any way desired. For demonstration purposes, Type 304 stainless steel was studied at 24 and 538°C. Sample microcomputer-generated crack growth plots of one specimen at 24°C are shown in Figs. 8 and 9 for d-c potential and COD methods, respectively. Note that the scatter for the d-c potential method is very small, while that for the COD method is much higher. This is primarily because the sampling frequency and system control was based upon the d-c potential resolution. Typical data generated at 24 and 538°C from several specimens are shown in Fig. 10 for the d-c potential method.

Conclusions

A microcomputer system has been developed to measure near-threshold fatigue crack growth. The system monitors crack length by both compliance and d-c potential methods, controls ΔK decay rate to a user-specified level, and calculates the crack growth rate as the test progresses. The crack lengths, loads, and temperatures are stored on disks for post-processing and plotting. The testing system has operated at temperatures up to 538°C.

References

[1] Bucci, R. J., "Development of a Proposed ASTM Standard Test Method for Near-Threshold Fatigue Crack Growth Rate Measurement," *Fatigue Crack Growth Measurement, ASTM STP 738*, American Society for Testing and Materials, Philadelphia, 1981, pp. 5–28.

[2] Saxena, A., Hudak, S. J., Jr., Donald, J. K., and Schmidt, D. W., *Journal of Testing and Evaluation*, Vol. 6, No. 3, May 1978, pp. 167–174.

[3] Carden, A. E., "A Critical Evaluation of Fatigue Crack Growth Measurement Techniques for Elevated Temperature Applications," WRC Bulletin 283, Welding Research Council, Feb. 1983.

[4] Wei, R. P. and Brazill, R. L., "An Assessment of A-C and D-C Potential Systems for Monitoring Fatigue Crack Growth," *Fatigue Crack Growth Measurement and Data Analysis, ASTM STP 738*, American Society for Testing and Materials, Philadelphia, 1981, pp. 103–119.

[5] Sullivan, A. M. and Crooker, T. W., "A Crack-Opening-Displacement Technique for Crack Length Measurement in Fatigue Crack Growth-Rate Testing Development and Evaluation," *Engineering Fracture Mechanics*, Vol. 9, 1977, pp. 159–166.

[6] Saxena, A. and Hudak, S. J., Jr., "Review and Extension of Compliance Information for Common Crack Growth Specimens," *International Journal of Fracture*, Vol. 14, No. 5, 1978, pp. 453–468.

[7] Yoder, G. R., Cooley, L. A., and Crooker, T. W., "Procedures for Precision Measurement of Fatigue Crack Growth Rate Using Crack-Opening Displacement Techniques," *Fatigue Crack Growth Measurement and Data Analysis, ASTM STP 738*, American Society for Testing and Materials, Philadelphia, 1981, pp. 85–102.

Peter K. Liaw,[1] William A. Logsdon,[1] Lewis D. Roth,[1] and Hans-Rudolf Hartmann[2]

Krak-Gages for Automated Fatigue Crack Growth Rate Testing: A Review

REFERENCE: Liaw, P. K., Logsdon, W. A., Roth, L. D., and Hartmann, H.-R., "Krak-Gages for Automated Fatigue Crack Growth Rate Testing: A Review," *Automated Test Methods for Fracture and Fatigue Crack Growth, ASTM STP 877,* W. H. Cullen, R. W. Landgraf, L. R. Kaisand, and J. H. Underwood, Eds., American Society for Testing and Materials, Philadelphia, 1985, pp. 177–196.

ABSTRACT: Bondable Krak-gages were utilized to monitor fatigue crack growth rates (FCGR) of various steels in air and salt water at 24°C and in wet hydrogen at 80°C. The FCGR data generated via bondable Krak-gages were consistent with those generated via either the compliance technique or optical measurements. Moreover, Krak-gages were effective in monitoring short crack extension.

Using atomic fusion technology, Krak-gages were sputtered on specimens so as to perform high-temperature (288 and 427°C) FCGR experiments. The FCGR results obtained with the sputtered Krak-gages were in good agreement with those obtained by the compliance method.

The Krak-gages were found to be a valuable addition to fracture mechanics testing. They readily interface with computers for fully automated FCGR experiments.

KEY WORDS: Krak-gage, Fractomat, fatigue crack growth, propagation, automated, computer, environment, sputter, compliance, atomic fusion, bondable, corrosive, air, fracture mechanics, data acquisition and analysis, short crack

Fracture mechanics has been a useful tool in life prediction and failure analysis of machine hardware. In order to accurately determine the fatigue life of a structural component, it is necessary to develop the fatigue crack growth rate (FCGR) properties of the materials of interest in their actual service environments. Due to the different configurations of machine components and their various operating conditions, fracture mechanics scientists

---

[1] Senior engineers and principal engineer, respectively, Metallurgy Department, Westinghouse R&D Center, Pittsburgh, PA 15235.
[2] President, Hartrun Corp., Chaska, MN 55318

are constantly searching for new techniques to monitor crack extension. In this paper, we have reviewed our experience in applying Krak-gages to FCGR testing. Over the past few years, bondable as well as sputtered Krak-gages have been used in several environments at the Westinghouse R&D Center. The Krak-gages were used to monitor both long and short fatigue crack growth behavior. The Krak-gages were interfaced with a computer for automated data acquisition and analysis.

Principle of the Krak-Gage

The bondable or sputtered Krak-gage is principally an indirect d-c potential measurement technique. Details of the Krak-gage theory were previously documented [1–4]. Following are brief descriptions of the bondable and sputtered Krak-gages.

Bondable Krak-Gage

The gage is a bondable, thin, electrically insulated metal foil of certain dimensions, photo-etched from a constantan alloy. The gage backing is a flexible epoxy-phenolic matrix providing the desired insulation and bonding surface area. By a suitable choice of gage size, gage material, adhesive bonding and backing, a crack will simultaneously propagate in both the bondable Krak-gage and the test specimen.

A constant current source of approximately 100 mA is employed to excite the low-resistance Krak-gage, Fig. 1a. A propagating crack produces a large change in the resistance of the gage and results in a high d-c output that is proportional to crack length. The output voltage (ΔV) of the gage, 0 to 100 mV, is further amplified to 10 V d-c full scale as shown in Fig. 1b. Overall system accuracy was found to be 2% or better of the gage full-scale rating, while the incremental crack extension (Δa) measurement was reported to be accurate within 0.1% [5]. The accompanying Fractomat instrumentation provides the necessary excitation and conditioning circuitry (Fig. 1b). The measured crack length data can be directly displayed on a digital voltmeter calibrated in millimetres. Furthermore, analog outputs are supplied to readily interface with all conventional recording instrumentation, data acquisition systems, and computers to allow automated crack propagation experiments. Figure 2 illustrates a 20-mm Krak-gage bonded to a compact-type specimen.

Sputtered Krak-Gage

Adhesives and organic insulating backings used in the bondable gage degrade in certain corrosive or high-temperature environments above 200 to 230°C. Atomic bonding, for example, sputtered coating, is more resistant to degradation in severe testing conditions. A gage atomically bonded to a test

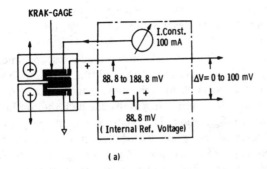

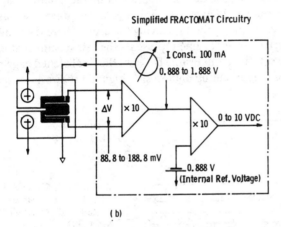

FIG. 1—*Krak-gage/Fractomat schematic.*

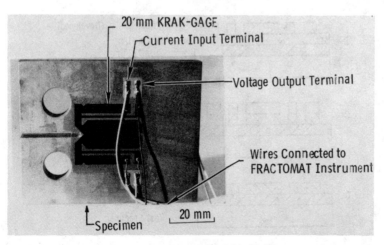

FIG. 2—*Specimen with bondable Krak-gage.*

specimen may be obtained by sputter-depositing a conductive coating on the specimen. The details of the development of the sputtered gage were previously reported [4].

Briefly, a thin quartz film was first sputtered on a test specimen to provide an insulating layer (Fig. 3). Secondly, a metal film compatible with the test material, such as an nickel-chromium alloy, was sputtered on the quartz layer. A photolithographic technique was used to etch the conductive coating into a desired gage configuration. The photo-etching process is illustrated in Fig. 3. Frame A presents the photosensitive resin, commonly referred to as photoresist, being exposed to ultraviolet light through a photographic mask of the gage pattern. Frame B illustrates the specimen after exposure to light and development, where the exposed areas of the photoresist remain and shield the desired conductive gage film areas. The unexposed areas are dissolved in the developing process, leaving the areas where the conductive coating is to be etched away. Frame C presents the test specimen after processing in the etching solution, and Frame D shows the final etched gage pattern after the removal of the remaining photoresist. Figure 4 shows a test specimen with

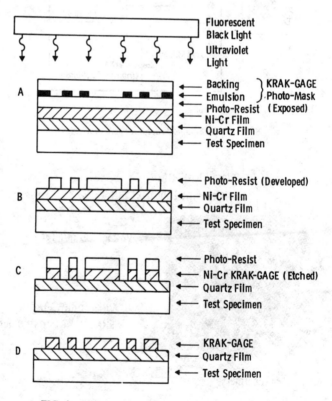

FIG. 3—*Photo-etching sequence of a sputtered Krak-gage.*

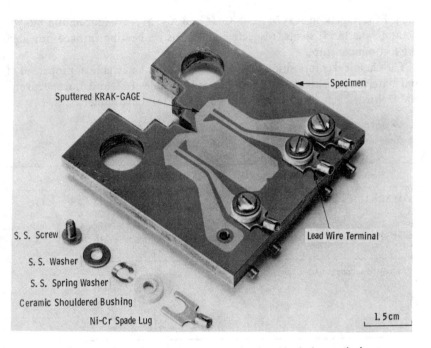

FIG. 4—*Specimen with sputtered Krak-gage and lead wire terminals.*

a sputtered Krak-gage and a specially designed lead wire system for high-temperature testing. It should be noted that the basic operating principle of the sputtered Krak-gage is the same as that of the bondable gage.

Experimental Procedure

Material and Environment

Bondable Krak-gages were used to develop FCGR data in several environments; that is, laboratory air and salt water at 24°C and wet hydrogen (plus 5% air) at 80°C. The materials investigated included an ASTM A508 Cl 2a automatic submerged-arc weldment, an HY 80 steel, and a manganese-chromium steel for air, salt water, and wet hydrogen environments, respectively. The saltwater environment contained 3.5% sodium chloride.

The short fatigue crack growth rate testing was conducted in the wet hydrogen (plus 5% air) environment. To provide the wet hydrogen environment, water was placed at the bottom of the chamber. The hydrogen gas pressure was maintained at 655 kPa. A heating tape wrapped around the chamber was used to maintain the desired temperature of 80°C.

Sputtered Krak-gages were utilized to perform 288 and 427°C air environment FCGR tests on ASTM A508 Cl 3a and ASTM A470 (chromium-molybdenum-vanadium) steels, respectively. A resistive heating furnace provided the test temperature.

The chemical compositions, heat treatments, and mechanical properties of the materials investigated are summarized in Tables 1, 2, and 3, respectively. Table 4 lists each material and the associated test environment.

TABLE 1—*Chemical compositions (weight %).*

Steel	C	Mn	P	S	Cu	Si	Ni	Cr	Mo	V	N
ASTM A508 Cl 2a[a]	0.11	0.62	0.006	0.011	...	0.33	0.60	<0.10	0.45	<0.12	...
HY 80	0.14	0.28	0.05	0.017	0.11	0.20	2.21	1.23	0.24	0.003	...
Mn-Cr	0.56	18.2	0.023	0.002	...	0.45	0.14	4.77	0.04	0.30	0.12
ASTM A508 Cl 3a	0.21	1.30	0.008	0.009	...	0.29	0.62	0.18	0.59	0.10	...
ASTM A470	0.30	0.80	0.009	0.002	...	0.20	0.30	1.20	1.20	0.25	...

[a]Automatic submerged-arc weldment.

TABLE 2—*Heat treatments.*

Steel	Heat Treatment
ASTM A508 Cl 2a[a]	austenitized at 860°C, water-quenched, tempered at 666°C, furnace cooled, stress-relieved at 607°C
HY 80	quenched from 899°C, tempered at 693°C
Mn-Cr	solution-treated, water-quenched, cold-expanded
ASTM A508 Cl 3a	austenitized at 871°C, water-quenched, tempered at 666°C, air-cooled
ASTM A470	austenitized at 950°C, air-cooled, tempered at 680°C

[a]Automatic submerged-arc weldment.

TABLE 3—*Mechanical properties at 24°C.*

Steel	0.2% Offset Yield Strength, MPa	Ultimate Tensile Strength, MPa
ASTM A508 Cl 2a[a]	601	669
HY 80	620	...
Mn-Cr	1120	1280
ASTM A508 Cl 3a	638	766
ASTM A470	662	775

[a]Automatic submerged-arc weldment.

TABLE 4—*Specimen dimensions and test conditions.*

| Steel | Environment | R-Ratio | KRAK-GAGE | Specimen Dimensions, mm | | | |
				Width	Height	Thickness	Notch Length
ASTM A508 Cl 2a	air, 24°C	0.2	bondable	51	61	6.4	13
HY 80	3.5% NaCl solution, 24°C	0.1	bondable	65	63	10	13
Mn-Cr	wet hydrogen (plus 5% air), 655 kPa, 80°C	0.1	bondable	51	61	6.4	...[a]
ASTM A508 Cl 3a	air, 288°C	0.2	sputtered	51	61	6.4	13
ASTM A470	air, 427°C	0.1	sputtered	51	61	13	13

[a]See Fig. 5.

Fatigue Crack Growth Testing

Crack propagation experiments were conducted using an automated MTS electrohydraulic fatigue machine [6]. Compact-type (CT) specimens were employed to generate FCGR data. Dimensions of the test specimens are presented in Table 4. Prior to testing, the CT specimens were precracked in accordance with the ASTM Test Method for Constant-Load-Amplitude Fatigue Crack Growth Rates Above 10^{-8} m/Cycle (E 647-81). The short crack growth experiment was performed on a CT specimen containing a blunt notch, Fig. 5. The short crack initiated from the root of the blunt notch. Details of the short crack work can be found in Ref 7.

The Krak-gages used in this investigation had a gage length of 20 mm. The gage was bonded slightly ahead of the notch tip toward the front face of the specimen. In some cases, Krak-gages were mounted on both sides of the test specimen.

At the beginning of each test, the output voltage of the gage was nulled by adjusting the calibration controls on the Fractomat instrument. A razor blade was then pressed into the tip of the specimen notch. Consequently, an initial reference reading was established on the digital display of the readout instrument. As the test progressed, the fatigue crack length was equal to the change of the reading (from the reference) plus the depth of the specimen notch.

In order to verify the crack length data developed by the Krak-gages, a clip gage mounted on the front face of the specimen was used to monitor crack length by a compliance technique [8]. Moreover, an optical microscope with a ×10 magnification was utilized to visually monitor crack extension. To confirm the short crack growth rate results obtained by the Krak-gage, striation spacings on the specimen fracture surface were carefully measured using scanning electron microscopy (SEM).

Test frequencies in the saltwater and wet hydrogen environments were 0.5

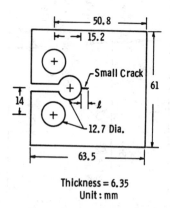

Thickness = 6.35
Unit : mm

FIG. 5—*Specimen geometry of short crack growth testing.*

and 0.02 Hz, respectively, while the air environment test frequency equaled 10 Hz. In this investigation a sinusoidal waveform was utilized at load ratios ($R = P_{min}/P_{max}$ where P_{min} and P_{max} were the applied minimum and maximum loads, respectively) of 0.1 and 0.2 (Table 4). Each FCGR test was conducted under a constant applied load range ($\Delta P = P_{max} - P_{min}$).

Automated Data Acquisition and Analysis

The MTS electrohydraulic fatigue machine interfaced with a PDP11/34A computer was used for the automated FCGR experiments. Details of the computerized FCGR test technique were previously reported [6]. A schematic of the computerized fatigue testing hardware is presented in Fig. 6. The analog outputs of the Krak-gage/Fractomat instrument and the clip gage were wired to the computer for automated data acquisition and analysis.

Signals from the load cell were used as feedback signals for crack propagation testing; thus the test machine was under load control. The command signals were developed by a function generator under direct program control using the PDP11/34A computer. The interface between the test stand and the computer consisted of analog-to-digital (A/D) conversion and timing circuitry, and a function generator. The analog-to-digital conversion was achieved by a 12-bit A/D converter. Using this computerized test system, increasing, decreasing and constant-ΔK fatigue crack propagation testing can be easily performed [6,8,9].

The data acquisition and analysis commands were expressed in BASIC language. Using the clip gage, crack length was measured by the compliance technique in which the load-versus-displacement data points were least-squares fitted by the computer [6]. The seven-point incremental polynomial method was utilized to convert crack length (a) versus elapsed cycle (N) to crack growth rate (da/dN) [10]. During the experiment, the graphics terminal of the computer displayed the loading cycles, applied loads, and the instantaneous da/dN-versus-ΔK results determined by either the Krak-gage or the clip gage. The visual measurements of crack length were periodically recorded in the computer for comparison with the results developed by the Krak-gage or clip gage.

Results and Discussion

Bondable Krak-Gage

Figures 7 and 8 present the crack length (a) versus loading cycle (N) data in air and saltwater environments at 24°C. The crack lengths determined by the Krak-gage, compliance, or visual technique are in excellent agreement in both environments. The FCGR data, da/dN versus ΔK, are shown in Figs. 9 and 10. The growth rates developed by the Krak-gage are identical to those

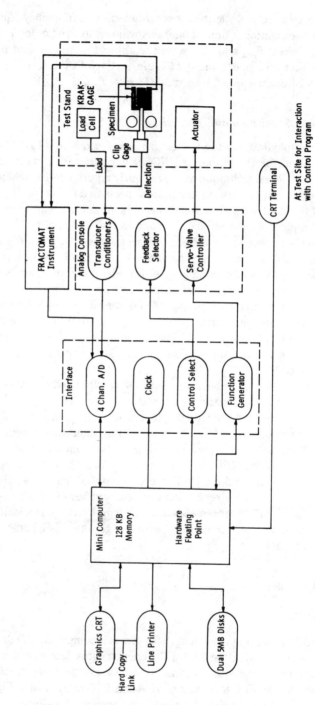

FIG. 6—Schematic of automated fatigue testing hardware.

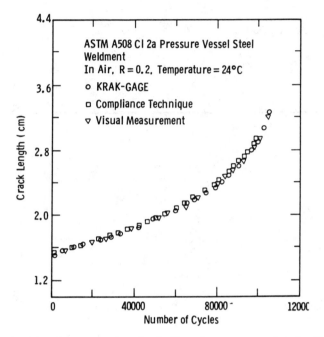

FIG. 7—*Crack length versus number-of-cycles data of a pressure vessel steel weldment in air at 24°C.*

generated via either the compliance or visual method. These results suggest that Krak-gages can be effectively used to monitor crack extension when developing the FCGR properties of various materials in air and saltwater environments.

The short FCGR results in 80°C wet hydrogen are illustrated in Fig. 11, where da/dN is plotted versus ℓ (crack length), the distance from the root of the blunt notch (Fig. 5). It was observed that the rates of short crack propagation initially decreased with increasing ℓ, then increased with a further increase in ℓ. Identical short crack growth characteristics were previously reported [11–18].

The fracture surface of the short crack growth specimen was carefully examined by SEM. The mode of fracture was transgranular with the presence of fatigue striations regardless of crack length (see Fig. 11). The striation spacing was found to decrease with increasing ℓ and then increase with further increasing ℓ. This behavior was consistent with small crack growth rate results determined by the Krak-gage. Moreover, striation spacings were measured and plotted against ℓ, as illustrated in Fig. 11. It was found that striation spacings were reasonably consistent with the crack growth rate data predicted by the Krak-gage. These results indicate that Krak-gages can be used to monitor small crack extension.

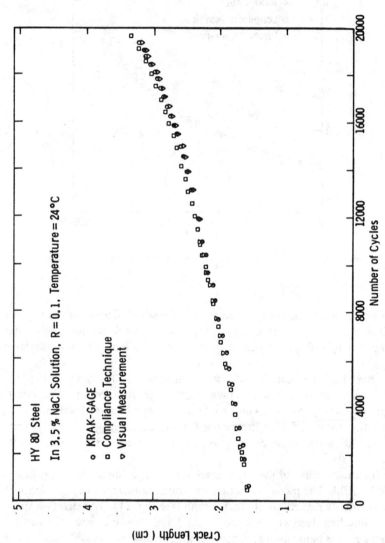

FIG. 8—*Crack length versus number-of-cycles data of HY80 steel in 3.5% sodium chloride solution at 24°C.*

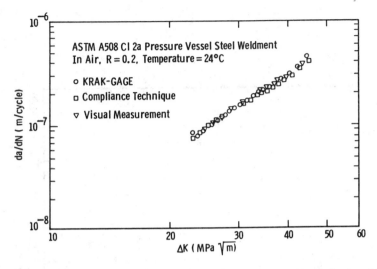

FIG. 9—*Fatigue crack growth rate data of a pressure vessel steel weldment in air at 24°C.*

As mentioned before, accuracy in the incremental crack length measurements by Krak-gages was reported to be better than 0.1% of the gage length. In the present short crack propagation investigation the gage length equaled 20 mm, which gave an accuracy level of better than 0.02 mm in measuring short crack extension. A proper choice of gage size is important in short crack growth studies. The smallest currently available Krak-gage has a gage length of 5 mm, which can easily detect a crack growth increment of less than 0.005 mm. Recently, Berchtold [19] employed a minicomputer-automated resonance-type test system to determine fatigue threshold stress-intensity range on small Charpy specimens using 5-mm Krak-gages. A system sensitivity of 0.000125 mm was reported to represent one minor digit with the 12-bit A/D converter. This further suggests a very high resolution of the Krak-gage for small crack growth testing into the fatigue threshold region of 2×10^{-11} m/cycle. Successful application of Krak-gages to near-threshold fatigue crack propagation testing was also confirmed by Jablonski [20].

Sputtered Krak-Gage

Results of crack length measurements using sputtered Krak-gages are presented in Figs. 12 and 13. For the 288°C test, Krak-gages were sputtered on both sides of the test specimen. Crack lengths measured by these two gages were in good agreement. Furthermore, crack length measurements by the Krak-gages and the compliance technique were consistent. At 427°C, the crack length data generated by the Krak-gage and the compliance method were comparable.

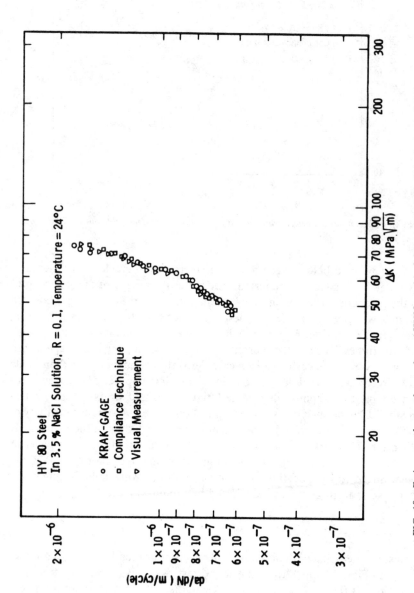

FIG. 10—*Fatigue crack growth rate data of HY80 steel in 3.5% sodium chloride solution at 24°C.*

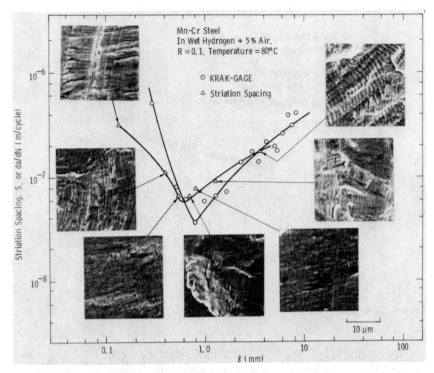

FIG. 11—*Comparison of crack growth rates determined by Krak-gage and striation spacing measurements in manganese-chromium steel.*

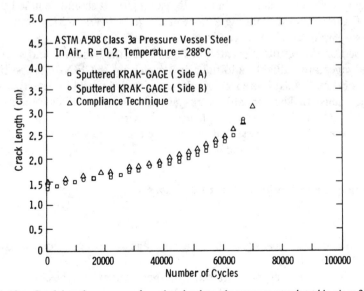

FIG. 12.—*Crack length versus number-of-cycles data of a pressure vessel steel in air at 288°C.*

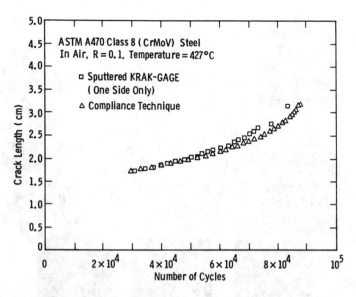

FIG. 13—*Crack length versus number-of-cycles data of a chromium-molybdenum-vanadium steel in air at 427°C.*

At 288 and 427°C, the FCGR results determined by the Krak-gages were reasonably consistent with those developed via the compliance technique; see Figs. 14 and 15, respectively. The maximum deviation in the FCGR measurements between both techniques was within a factor of 1.5, which was less than the typical scatter factor of 2 in FCGR testing [*10*]. It should be noted that at 288°C there was excellent agreement in the da/dN-versus-ΔK data between the two Krak-gages (see Fig. 14).

Based on the results presented in Figs. 7–15, the bondable and sputtered Krak-gages are valuable additions in developing the FCGR properties of structural materials. Moreover, Krak-gages proved useful in detecting small crack extension. The bondable Krak-gages performed well in air and salt water at 24°C and in an 80°C wet hydrogen environment while the sputtered Krak-gages proved successful at monitoring crack extension at the relatively high temperatures of 288 and 427°C.

Advantages and Limitations of Krak-Gages

Advantages

1. Krak-gages can easily replace the clip gage for fully computer-controlled FCGR testing [*6,9*].

2. It is feasible that the sputtered Krak-gage will extend the effective range of applicability of this technology to 1000°C. Furthermore, it is suggested

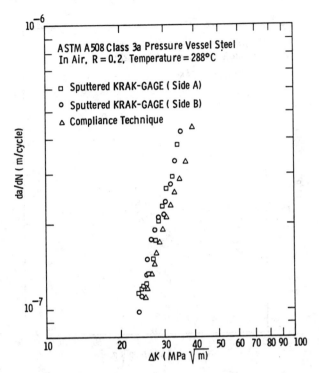

FIG. 14—*Fatigue crack growth rate data of a pressure vessel steel in air at 288°C.*

that the sputtered gage may be applicable to cryogenic temperature fracture mechanics testing.

3. It is expected that Krak-gages can be directly sputtered onto nonconductive materials such as glass, ceramics, and plastics for fracture-type testing at very high crack velocities approaching 3000 m/s [21].

4. The geometry of the Krak-gage can be easily altered to accommodate the configuration of a structural component. Recently, a gage with a 200-mm crack length was developed for automated FCGR testing on 1219-mm-wide center-notched airframe test panels.

Limitations

1. Unfortunately, Krak-gages can only be utilized to measure surface crack lengths. If severe crack tunneling is experienced in some type of fracture mechanics testing (for example, stress-corrosion cracking experiments), Krak-gages are not suitable for monitoring crack length.

2. The duration capability of Krak-gages subjected to highly corrosive environments is still unknown. On one occasion, sputtered gages were employed

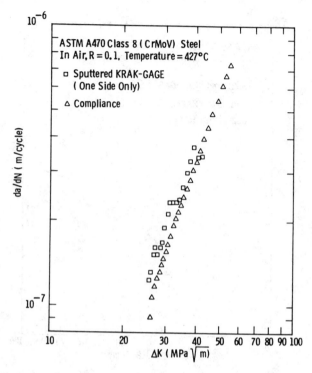

FIG. 15—*Fatigue crack growth rate data of a chromium-molybdenum-vanadium steel in air at 427°C.*

to develop the FCGR properties of pressure vessel steels in a pressurized water environment (temperature 288°C, pressure 13.8 MPa). The test frequency was slow, that is, 1 cpm, such that the FCGR tests often lasted from one to two months. By the end of the experiments the sputtered gages were found to be pitted and nonoperative. Therefore, further research is needed to refine the sputtering technique such that improved Krak-gages can be made available for long-term corrosion FCGR testing.

Conclusion

Bondable Krak-gages have been successfully utilized in determining FCGR properties in ambient air and corrosive environments. The crack growth rate data developed by Krak-gages proved to be consistent with those developed via either the compliance technique or visual measurements. Furthermore, the rates of short crack growth can be monitored by these gages. Moreover, the sputtered Krak-gage proved to be useful in developing high-temperature FCGR data.

The development of the Krak-gage provides a valuable addition to fracture mechanics testing. The gage lends itself readily to automated data acquisition and test analysis and, therefore, can save costly labor associated with FCGR testing. Moreover, Krak-gages can be interfaced with a computer for direct machine control in performing fully automated FCGR experiments.

Acknowledgments

We are very grateful to Dr. A. Saxena for many constructive suggestions and for allowing us to use the crack growth rate results in Fig. 11. We wish to thank A. R. Petrush, M. G. Peck, R. A. Hilgert, R. S. Williams, E. J. Helm, P. J. Barsotti, and G. McFetridge for their assistance in conducting the experiments. The FCGR tests were conducted in the Mechanical Behavior Laboratory at the Westinghouse R&D Center under the direction of R. E. Gainer. We are additionally grateful to R. Stuart of the Koral Laboratories for sputtering the Krak-gages.

References

[1] Hartmann, H. R. and Churchill, R. W., "KRAK-GAGE: A New Transducer for Crack Growth Measurement," presented at the Society for Experimental Stress Analysis Fall Meeting, sponsored by the Society for Experimental Stress Analysis, Brookfield Center, CT, Oct. 1981.

[2] Paris, P. C. and Hayden, B. R., "A New System for Fatigue Crack Growth Measurement and Control," presented at the Symposium on Fatigue Crack Growth, sponsored by the American Society for Testing and Materials, Pittsburgh, PA, Oct. 1979.

[3] Liaw, P. K., Hartmann, H. R., and Helm, E. J., *Engineering Fracture Mechanics*, Vol. 18, 1983, p. 121.

[4] Liaw, P. K., Hartmann, H. R., and Logsdon, W. A., *Journal of Testing and Evaluation*, Vol. 11, 1983, p. 202.

[5] "Krak-Gage Accuracy and Resolution," KRAK-TIP No. 8109-1, TTI Division, Hartrun Corp., Chaska, MN, 1981.

[6] Williams, R. S., Liaw, P. K., Peck, M. G., and Leax, T. R., *Engineering Fracture Mechanics*, Vol. 18, 1983, p. 953.

[7] Saxena, A., Wilson, W. K., Roth, L. D., and Liaw, P. K., "The Behavior of Small Fatigue Cracks at Notches in Corrosive Environments," *International Journal of Fracture*, in press.

[8] Saxena, A. and Hudak, S. J., Jr., *International Journal of Fracture*, Vol. 14, 1978, p. 453.

[9] Saxena, A., Hudak, S. J., Jr., Donald, J. K., and Schmidt, D. W., *Journal of Testing and Evaluation*, Vol. 6, 1978, p. 167.

[10] Clark, W. G., Jr. and Hudak, S. J., Jr., *Journal of Testing and Evaluation*, Vol. 3, 1975, p. 454.

[11] Pearson, S., *Engineering Fracture Mechanics*, Vol. 7, 1975, p. 235.

[12] Lankford, J., *Fatigue of Engineering Materials and Structures*, Vol. 5, 1982, p. 233.

[13] James, M. R. and Morris, W. L., *Metallurgical Transactions*, Vol. 13A, 1982, p. 153.

[14] El Haddad, M. H., Smith, K. N., and Topper, T. H., *Journal of Engineering Materials and Technology, Transactions*, American Society of Mechanical Engineers, Series H, Vol. 101, 1979, p. 42.

[15] Tanaka, K. and Nakai, Y., *Fatigue of Engineering Materials and Structures*, Vol. 6, 1983, p. 315.

[16] Suresh, S. and Ritchie, R. O., "The Propagation of Short Fatigue Cracks," *International Metals Reviews*, Vol. 29, 1984, p. 445.

[17] Hudak, S. J., Jr., *Journal of Engineering Materials and Technology, Transactions*, American Society of Mechanical Engineers, Series H, Vol. 103, 1981, p. 26.

[*18*] Liaw, P. K. and Logsdon, W. A., "Crack Closure: An Explanation for Small Fatigue Crack Growth Behavior," *Engineering Fracture Mechanics* (in press).

[*19*] Berchtold, R., "Computer Controlled Fatigue Testing Machine for Determination of Stress Intensity ΔK_{th} at Threshold," Report No. 8303.02-3/1, Russenberger AG, Schaffhausen, Switzerland, presented at City University, London, March 1983.

[*20*] Jablonski, D. A., pp. 269–297.

[*21*] Kobayashi, A. S., Emery, A. F., and Liaw, B. M., *Journal of the American Ceramic Society*, Vol. 66, 1983, p. 151.

Gin Lay Tjoa,[1] *François P. van den Broek,*[1] *and Bart A. J. Schaap*[1]

Automated Test Methods for Fatigue Crack Growth and Fracture Toughness Tests on Irradiated Stainless Steels at High Temperature

REFERENCE: Tjoa, G. L., van den Broek, F. P., and Schaap, B. A. J., "Automated Test Methods for Fatigue Crack Growth and Fracture Toughness Tests on Irradiated Stainless Steels at High Temperature," *Automated Test Methods for Fracture and Fatigue Crack Growth, ASTM STP 877,* W. H. Cullen, R. W. Landgraf, L. R. Kaisand, and J. H. Underwood, Eds., American Society for Testing and Materials, Philadelphia, 1985, pp. 197-212.

ABSTRACT: An automated system for fatigue crack growth and fracture toughness measurements has been developed for irradiated stainless steels tested at temperatures up to 925 K. The system, including a microcomputer, is based on the d-c potential-drop technique for crack extension measurements. Specimens of the compact-tension type are used for the experiments. A description is given of the potential-drop method and the automated data acquisition system. The method of collecting *N-a* data pairs is given as well as the calculation and analysis of the fatigue crack growth rate (da/dN) and stress-intensity factor (ΔK). The measurement of load, deflection, and crack extension data to determine the *J*-versus-Δa curves is also discussed.

Calculations and interpretations of the results are in good agreement with the ASTM standards.

The system for crack growth experiments has proven to be very reliable, with a high resolution and accuracy.

In practice, good experience has been gained using the system for testing irradiated specimens under remote handling conditions.

KEY WORDS: fatigue, fatigue crack growth, fracture toughness, stainless steel, electric potential, data analysis, data acquisition system, irradiation, high temperature

[1]The Netherlands Energy Research Foundation, ECN, Materials Department, Petten (NH), The Netherlands.

For several years fatigue crack growth and fracture mechanics experiments have been performed at the Materials Department of the Netherlands Energy Research Foundation (ECN).

The tests are conducted on irradiated specimens at temperatures up to 925 K, using a heat-resistance split furnace, allowing the d-c leads to come out of the furnace, which is provided with a quartz window for optical observation of the crack.

The data are used in the design and safety analysis of Liquid Metal Fast Breeder Reactor (LMFBR) primary components, such as reactor vessel and grid plate. Further, the experiments provide information contributing to the understanding of physical mechanisms affecting the fatigue life.

Continuous monitoring of crack length measurements is required because the test duration often exceeds working hours. It is achieved using a d-c potential-drop technique. This technique has gained acceptance as a reliable, accurate, and cost-effective method for measuring crack lengths in fatigue specimens [1,2].

The d-c equipment has been integrated into a microprocessor-based data acquisition system to provide for the required data processing and analysis. The results are presented in tables and graphs.

A description of the d-c potential-drop system and the equipment to automate it is given herein. The measuring method as well as some results are also discussed.

Equipment and Measuring Method

D-C Potential Drop System

A d-c potential-drop system was adopted by the Mechanical Testing Group, ECN Materials Department.

Since irradiated compact-tension (CT) specimens (Fig. 1), with and without attachment points for the potential leads, have to be tested, a suitable joining method applicable for hot-cell handling had to be developed. The connection of the leads to the specimen is achieved by screwing threaded ends into the specimen. The leads are also provided with brazed nuts for remote handling purposes.

The holes in the specimens are drilled using a template, thus allowing a good reproducibility of the contact area and contact resistance. The current and potential leads are both 3 mm in diameter.

The potential differences, which are measured from a test specimen, are dependent upon the magnitude of the current, specimen shape and size, specimen resistivity, and the positions of the current leads and potential probes.

Performing tests in a hot cell compels, irrespective of any measuring method, certain restrictions. A main restriction is, for instance, accessibility to the setup using optical devices at high temperature. Since a hot cell is a

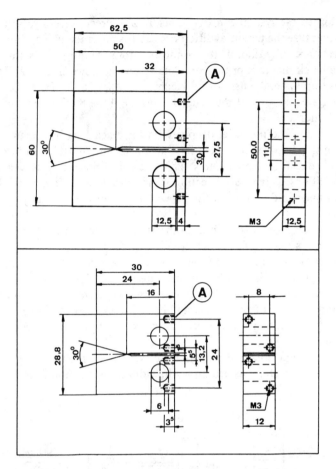

FIG. 1—*Dimensions in mm of compact-tension specimen used for fatigue crack growth exper-iments* (above) *and fracture toughness tests* (below) *showing current lead and potential probe attachment points* (A).

shielded and conditioned facility, the temperature is very stable. Therefore, no influence of the environment temperature on the d-c potential will be observed. From the literature it is known that a welded joint for the d-c leads is most preferable. However, a good reproducibility of the positions of the leads under remote handling conditions cannot be guaranteed. Therefore, the position of the leads is a compromise between the most favorable position as known from experience and the literature [1,2] and the requirements in connection with practical realization in hot cells. Additional care is given with respect to the threaded connection that a maximum metallic contact is obtained. The junctions from potential leads to measuring instruments are situated outside the furnace and are easy accessible.

The adopted d-c system consists basically of a direct-current supply and a device to measure the potential differences across the crack plane. Since these components are fully digital, the connection to a computing system is very easy. The block diagram of the complete system is shown in Fig. 2. The d-c potential-drop system consists of a Kepco d-c supply unit, Type JQE 25-40 M(Y), and a Fluke digital multimeter (DMM), Type 8502 A, including an IEEE 488 interface.

Prior to using the potential-drop system for crack length measurements, the relationship between d-c potential and crack length must be calibrated, at the respective testing temperatures. This was done under static as well as dynamic conditions. During static calibrations, without applying loads, the d-c signal was measured on CT specimens with different mechanically prepared notches.

During crack growth (dynamic calibrations), optical crack length increments on both sides of the surface are measured.

A frequency marking technique is used to determine the actual crack length. In this way it is possible to correlate the d-c signal with the average crack length, including crack front curvature, to determine the relationship.

Although no "crack front tunneling" has been observed during our fatigue crack growth experiments, this method can also be applied to correct for such a phenomenon.

Calibration data obtained with the d-c system are shown in Fig. 3. The resolution of the system according to the calibration was found to be better

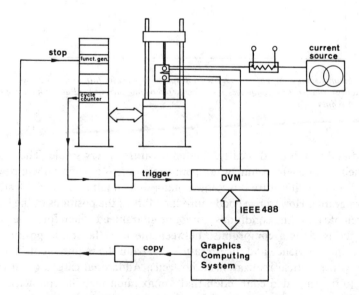

FIG. 2—*Block diagram of the data-acquisition system for fatigue crack growth including d-c potential drop system and machine control unit.*

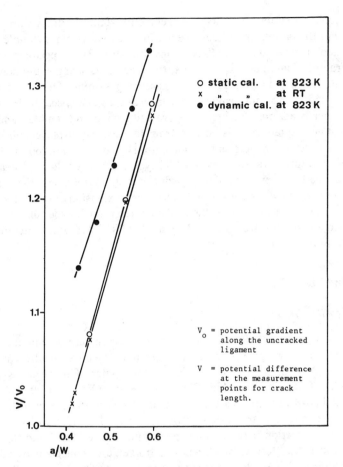

FIG. 3—*Results of static and dynamic calibrations on CT specimens.*

than 0.05 mm. The sensitivity ranges from 48 μV/mm at room temperature to 62 μ/mm at 823 K.

Data Acquisition

A fatigue crack growth experiment, especially at low frequencies, is time-consuming and many have to be continued without interruptions for several weeks. Therefore, a microcomputer-based data acquisition system was developed.

Essential for application of this system is the high stability and reliability of the d-c current supply and the accuracy of the digital voltmeter.

Apart from the 8000-type and 1251-type servohydraulic fatigue testing machines manufacturing by Instron and the d-c potential-drop equipment, the data acquisition system consists of a Tektronix 4052 Graphics Computing System. For the fatigue crack growth experiments this system is completed by a homemade trigger unit as well as a copy unit (Fig. 2). To perform elastic-plastic fracture toughness experiments, the crack measuring equipment was combined with a load and deflection measuring system. The deflection is measured with a ±5-mm displacement transducer. The specially developed extensometer is adapted for high-temperature testing and remote handling. The load-line displacement is derived from the relative displacement of the clevises. The clevis displacement is transmitted by an extension rod to a linear variable differential transformer (LVDT) type transducer, located near the bottom of the lower pullrod. The transducer is situated outside the furnace and is thermostatically water-cooled. The crack extension is measured with the same equipment as used for the fatigue crack growth experiments. Load and deflection signals are measured by an HP 3497 A scanner including a 5½-digit digital voltmeter (DVM).

Measuring Method

Fatigue Crack Growth

A constant current of 10 A is supplied by the current supply unit and measured accurately through a shunt. Measurement of the potential drop in a CT specimen is done directly by a 6½-digit d-c voltmeter. This device is interfaced with a TEK 4052 Graphics Computing System by means of the IEEE 488 bus. All DMM functions are addressable on a software level over this bus.

The crack length is measured at predetermined numbers of cycles by means of a connection between the trigger input of the DMM and the cycle counter unit of the testing machine. The trigger unit collects a transistor-transistor-logic (TTL) pulse from the testing system. This galvanically separated pulse is transmitted to frequency dividers, resulting in the respective outputs of the trigger unit after 1, 10, or 100 cycles.

The data logging frequency can thus be adjusted for the particular frequencies ranging from 0.01 to 100 Hz. An option to stop the testing machine at a predetermined crack length is also operational. Signal averaging is done to avoid discontinuities from noise, slight drifts of the signal due to the load cycle, or time response effects of the equipment.

The N/a curve can be visualized instantaneously. After each experiment da/dN-versus-ΔK values are calculated and the data are analyzed by regression analysis.

Fracture Toughness Tests

The fatigue precracking data are measured using the same equipment as for the crack growth experiments. For ductile crack growth, calibration procedures are used based on multispecimen technique and actual crack length measurements.

The load and deflection signals are scanned by an HP 3497 A. Both the scanner and DVM are connected to the TEK 4052 microcomputer (Fig. 4). In this way the relevant data to calculate the area under the load/loadline displacement curve are continuously available.

After processing, these data are stored in the microcomputer.

Data Processing and Analysis

Fatigue Crack Growth

The total software package consists of three programs: FATCRAG (fatigue crack growth), CRAGRAN (crack growth analysis), and CRACKSORT (crack growth sorter). A short description of these programs is now given as well as some results.

FATCRAG

The number of cycles (N) and the crack length (a) are stored using this program. The determination of crack length is based on the output of the d-c potential and calibration data. As previously mentioned, the sampling frequency is dependent on the test frequency. If an increase of crack length of

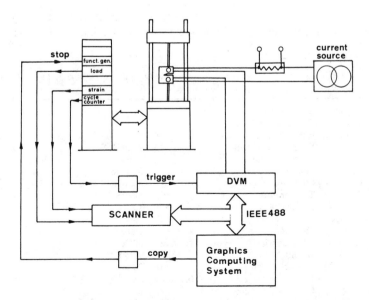

FIG. 4—*Block diagram of data-acquisition system for fracture toughness tests.*

0.1 mm is found, the match number of cycles and crack length is stored. Instantaneously a curve showing N versus a can be displayed.

CRAGRAN

Prior to the computing and data analysis, the validity of the crack length measurements is determined from the final crack length as optically measured on the fracture surface of the tested specimen. If necessary, the differences found between this measurement and the values determined by the d-c potential-drop system, using FATCRAG, can be corrected for.

New N-a data pairs then appear and are used to calculate da/dN-versus-ΔK values, for which the ΔK calculations are done according to the ASTM Test Method for Constant-Load-Amplitude Fatigue Crack Growth Rates Above 10^{-8} m/Cycle (E 647-83). The crack growth rate calculation is performed according the seven-point polynomial method described in Ref 3.

An output of the collected data and the calculated values is given in Tables 1–3.

Next, statistical analyses are performed, based on the Paris formula using linear regression analysis [4,5].

In the end, three graphic presentations are given, showing the N-versus-a curve, da/dN-versus-ΔK curve, and a curve of the regression line including the 95% confidence band (Figs. 5–7).

CRACKSORT

The data of every test are stored in data files. A preselection is done for test temperature and material condition (reference or irradiated). By this data

TABLE 1—*Printout of specimen identification and test parameters by CRAGRAN.*

Cassette code	:	F1
Data file number	:	36
Number of data pairs	:	100
Date of experiment	:	21 0KT.1981
Specimen code	:	C25
Material	:	SS type 304
Condition	:	irradiated plate
Temperature	(K):	823
Frequency	(Hz):	10
P-max, max load	(kN):	4.7
R-ratio	:	0.05
W-value width	(mm):	50.00
B-value thickness	(mm):	12.50
a_0-crack length	(mm):	20.50
Dc-calibration value	(μV/0.1 mm):	6.5
U_0-DC start value	(uV):	0
Measured interval	(10 100):	100

TABLE 2—*Printout of measured and calculated data by CRAGRAN.*

Cycles	Crack Length Measured, mm	Crack Length Calculated, mm	da/dN, mm/cycle	ΔK, MNm$^{-3/2}$
2 720	20.60			
6 420	20.70			
9 420	20.81			
12 720	20.90	20.90	3.2E-005	12.2
16 120	21.00	21.01	3.3E-005	12.3
18 820	21.10	21.10	3.5E-005	12.3
21 820	21.20	21.21	3.7E-005	12.4
24 620	21.32	21.31	3.8E-005	12.5
27 020	21.41	21.40	3.8E-005	12.5
29 520	21.50	21.50	3.9E-005	12.6
32 220	21.60	21.61	4.0E-005	12.7
34 820	21.71	21.71	4.1E-005	12.7
37 120	21.81	21.81	4.1E-005	12.8
39 420	21.91	21.90	4.2E-005	12.9
41 720	22.00	22.00	4.2E-005	13.0
128 120	28.91	28.92	2.4E-004	20.1
128 420	29.00	28.99	2.4E-004	20.2
128 920	29.11	29.11	2.4E-004	20.4
129 420	29.22	29.23	2.4E-004	20.6
129 720	29.31	29.29	2.4E-004	20.7
130 220	29.41	29.42	2.5E-004	20.9
130 620	29.51	29.52	2.6E-004	21.0
131 020	29.64	29.62	2.6E-004	21.2
131 320	29.71	29.71	2.7E-004	21.3
131 720	29.80	29.81	2.7E-004	21.5
132 120	29.92	29.92	2.9E-004	21.7
132 420	30.01	30.01	3.0E-004	21.8
132 820	30.12	30.13	3.1E-004	22.0
133 020	30.21	30.20	3.2E-004	22.2
133 420	30.33			
133 720	30.42			
134 020	30.54			

reduction procedure an efficient use of the magnetic storage tapes is obtained. Within the preselected collection a sorting operation can be performed on alloy, test frequency, or R-ratio.

Fracture Toughness

Prior to the monotonic loading test, the specimen is fatigue precracked. Based on the FATCRAG program, the final fatigue crack length is determined. Afterwards, monotonic loading is performed under ram displacement control at a constant displacement rate. Data collection, computation, and analyses are done with the program JTENS (tension test prior to J determination) and JCALC (calculation of J), respectively.

TABLE 3—*Results of Paris and statistical analysis as calculated by CRAGRAN.*

Paris Analysis:
    Constant: 2.439861252E-9
    Exponent: 3.79769380344
    Equation:
        $da/dN = 2.4E\text{-}009* (K_{max} - K_{min})^{3.8}$
        da/dN in mm/cycle
        ΔK in MNm$^{-3/2}$

Statistical Analysis:
    Number of analyzed data pairs   :   94
    Domain of data ΔK_{min} :   12.2
                   ΔK_{max} :   22.2
    Center of data ΔK :   16.4
                   da/dN :   1.0E-004
    Coefficient of determination   :   0.994930716336
    Estimated standard error   :   0.0282620011888
    Vertical half-width of 95% confidence: 0.0414739768003

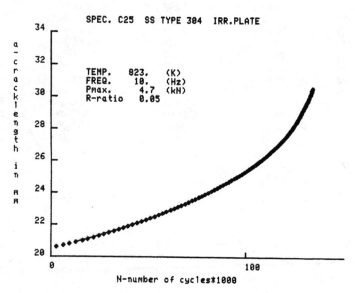

FIG. 5—*Graphic presentation of* a *versus* N *as performed by CRAGRAN.*

JTENS

The load-deflection data logging is based on the requested total deflection and total testing time, resulting in 100 data pairs for 1-mm deflection. Crack extension data are collected from the potential system, based on a detection limit of 0.1 mm. Thus every stored value will be a multiple of 0.1 mm. The

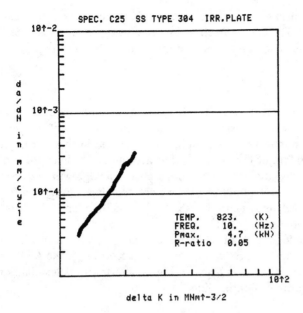

FIG. 6—da/dN-*versus*-ΔK *plot derived from* a/N *data.*

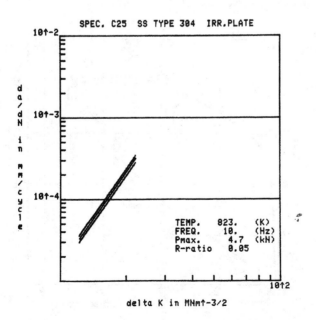

FIG. 7—*Calculated regression line with upper and lower confidence band.*

crack extension data are matched with the corresponding load-deflection data pairs. The load-deflection data and the crack extension-deflection data are graphically presented in combination, as shown in Fig. 8.

JCALC

For the early stage of the test the deflection values are corrected to load-line displacement values using a linear regression analysis of the early deflection data and the theoreticaly compliance of the specimen according to the measured final fatigue crack length. The area under the load/load-line displacement curve is calculated for the matched crack growth and load-deflection data pairs. Corresponding J-values are calculated using the formula for a growing crack as given in the ASTM Test Method for J_{Ic}, a Measure of Fracture Toughness (E 813-81).

The results are given in tabulated and graphical form. Then the crack growth resistance curve can be analyzed according to the ASTM recommendations. An example is given in Fig. 9 showing J_Q determination using the actual measured blunting line.

Discussion

The static calibration method using a sawcut to represent a crack can lead to an underestimation of the crack length. This is shown in our figures and recently discussed by Wilson [6].

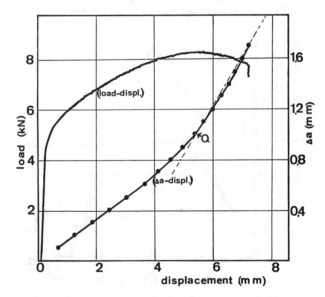

FIG. 8—*Computer plot of the load-deflection diagram and the crack extension (potential-drop) deflection diagram.*

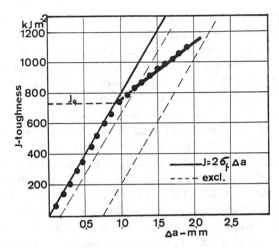

FIG. 9—*Single-specimen potential-drop J R curve.*

On the other hand, calibrations based on optical crack length increments can underestimate the true crack length when crack front curvature is not taken into account. By using marking techniques after each increment of crack growth, the exact average crack lengths can be determined. For the very first stage of the development of the fatigue crack, the linear relationship (within the applied a/W range) between V/V_0 and a/W does not exist. The crack length data are based on "original" $a_0 + \Delta a$ where Δa is defined by the constant calibration value from the assumed linear V/V_0-versus-a/W relationship for narrow a/W ranges.

The development of crack front curvature is partly taken into account in the testing procedure by starting the Δa measurements after a crack extension of 1 mm is reached.

Corrections for crack front curvature can be applied. Final crack length is measured optically on the fracture surface after completion of the test. From this measurement corrections for crack front curvature can be obtained.

The ΔK calculations are carried out using the equation mentioned in ASTM E 647-83. The accuracy of the CT specimen dimensions is within the specification of this standard. The relative error in ΔP is the sum of the static load calibration accuracy and that due to the dynamic load response of the testing machine. The total possible error in ΔP amounts to 0.3%. To estimate the accuracy of the calculated ΔK-value, an additional factor has to be taken into account, namely, the inaccuracy of the crack length measuring system.

The total possible error of specimen W and clevis holes adds up to about 0.6% W and about 0.1% for the crack length a. Further, it is evident that the as-fabricated specimen dimensions are important. If this procedure is followed, the total error in ΔK can be reduced by a factor 2.

The difference between deflection and load-line displacement can cause an error of about 100% for the area calculations during the early stage of the monotonic loading load-deflection curve. Therefore, the correction procedure is incorporated in the JCALC program. A selection criterion for using corrected load-line displacement values was based on the criterion of 5% difference between areas under the noncorrected and corrected curves. As soon as the difference is reduced below 5%, this correction, which can be ultimately applied up to maximum load, is ignored. This 5% criterion was chosen taking into account the ASTM accepted standard deviation of 17.4.

Other sources of error (extensively discussed in Refs *1* and *2* for which precautions were taken are secondary current paths, thermal effects, crack closure, and plasticity. To avoid random disturbance, caused by secondary currents, it is necessary to insulate the specimen from the loading device or machine frame. Insulation is achieved by using a trifluoroethylene plate between a pull rod and load cell. Thermal effects can be reduced by using potential probes of the same alloy as the tested specimen. At high temperature, symmetrical location of specimens and probes in the furnace as well as good temperature stability can also minimize this effect.

Crack closure can cause fluctuations in potential readings due to decreasing resistance over the remaining ligament. Using a computer-controlled data-acquisition system the potential measurement can be performed at maximum load, thus reducing the effect of crack closure. Plasticity is not important for linear elastic fracture mechanics crack growth experiments since plasticity is limited to a very small area ahead of the crack tip. Consequently, no corrections are necessary for fatigue crack growth experiments. However,

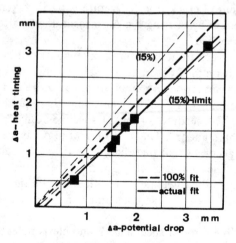

FIG. 10—*Comparison of crack extension values measured with the potential-drop method and the heat-tinting technique.*

if there is considerable plasticity as in the case of the elastic-plastic fracture toughness tests (J-tests), then these effects do contribute to the potential-drop signal. In that case it is always necessary to check the final crack length on the fracture surface and to correct for plasticity effects by empirical calibration based on data from multispecimen tests with different crack lengths, as shown in Fig. 10.

Results of experimental programs using the equipment and procedures mentioned in this paper were reported recently [7,8]. Irradiation effects on the post-irradiation fatigue crack growth and fracture toughness properties were measured with a high resolution thanks to the accuracy and reliability of the measuring technique and data analyzing procedures. In addition, the testing method has proven to be very useful and easy to handle for remote testing of irradiated specimens. The efficiency can be further explored when the same standard specimens (½T and 1T CT type) are used for different experimental programs.

Conclusions

1. The system for crack growth experiments has proven to be very reliable, with a high resolution and accuracy. The resolution of the system according to the calibration results was found to be better than 0.05 mm.

2. The automated data system is fast, and standardized data analysis is performed with high accuracy and reliability. Calculations and interpretations of the fatigue crack growth results are in good agreement with ASTM Method E 647-83.

3. The calibration figures given in this paper apply only to the given specimen geometries, current lead and probe wire configurations, and actual alloy, and are not valid in general.

4. Effects of irradiation on the testing parameters can be measured with a high resolution, thus improving the database and knowledge of the observed phenomena.

5. In practice this system has been shown to work satisfactorily for testing irradiated specimens at high temperature under remote handling conditions.

References

[1] The Measurement of Crack Length and Shape During Fracture and Fatigue, C. J. Beevers, Ed., Engineering Materials Advisory Services Ltd., Warley, West Midlands, U.K., 1980.
[2] Advances in Crack Length Measurement, C. J. Beevers, Ed., Warley, West Midlands, U.K., 1982.
[3] Clark, W. G., Jr., and Hudak, S. J., Jr., "Variability in Fatigue Crack Growth Rate Testing," Journal of Testing and Evaluation, Vol. 3, 1975, pp. 454-476.
[4] Paris, P. C. and Erdogan, F., "A Critical Analysis of Crack Propagation Laws," Journal of Basic Engineering, Series D, Transactions of the ASME, JBAEA, Dec. 1963, pp. 528-534.
[5] Fong, J. T. and Dowling, N. E. in Fatigue Crack Growth Measurement and Data Analysis, ASTM STP 738, S. J. Hudak, Jr., and R. J. Bucci, Eds., American Society for Testing and Materials, Philadelphia, 1981, pp. 171-193.

[6] Wilson, W. K., "On the Electrical Potential Analysis of a Cracked Fracture Mechanics Test Specimen using the Finite Element Method," *Engineering Fracture Mechanics*, Vol. 18, No. 2, 1983, pp. 349-358.

[7] de Vries, M. I. in *Effects of Radiation on Materials, ASTM STP 782*, H. R. Brager and J. S. Perrin, Eds., American Society for Testing and Materials, Philadelphia, 1982, pp. 720-734.

[8] de Vries, M. I. and Schaap, B. A. J., *Elastic-Plastic Fracture Methods: The User's Experience, ASTM STP 856*, American Society for Testing and Materials, Philadelphia, 1985, pp. 183-195.

Yi-Wen Cheng[1] and David T. Read[1]

An Automated Fatigue Crack Growth Rate Test System

REFERENCE: Cheng, Y.-W. and Read, D. T., **"An Automated Fatigue Crack Growth Rate Test System,"** *Automated Test Methods for Fracture and Fatigue Crack Growth, ASTM STP 877,* W. H. Cullen, R. W. Landgraf, L. R. Kaisand, and J. H. Underwood, Eds., American Society for Testing and Materials, Philadelphia, 1985, pp. 213–223.

ABSTRACT: An automated fatigue crack growth rate (FCGR) test system has been developed that can be used for tests of constant-load-amplitude FCGR above 10^{-8} m/cycle [ASTM Test Method for Constant-Load-Amplitude Fatigue Crack Growth Rates Above 10^{-8} m/Cycle (E 647-83)] at normal (~ 10 Hz) or low (~ 0.1 Hz) cyclic frequencies and for tests of near-threshold and variable-load-amplitude FCGR. The test system consists of a minicomputer, a programmable arbitrary waveform generator, a servo-hydraulic test frame, and a programmable digital oscilloscope. The crack length is measured using the compliance technique; the FCGR and the stress-intensity factor range are calculated and plotted automatically during the test.

KEY WORDS: automated test system, compliance technique, fatigue crack growth rate, fatigue of materials, near-threshold fatigue test, variable-load-amplitude fatigue test

Fatigue crack growth rate (FCGR) data are used for material characterization and for fracture mechanics reliability analysis of structures subjected to cyclic loading. A standard test method for measuring such data above 10^{-8} m/cycle under constant-amplitude loading has been developed and published in the *1983 Annual Book of ASTM Standards* under the designation ASTM E 647-83.

With the increased interest in near-threshold FCGR [1,2] and FCGR under environmental influences at low cyclic frequencies [3], the demand for FCGR measurements has increased. Obtaining such data can be tedious and time-consuming. An automated FCGR test system, such as that described in this paper, allows testing to proceed, data to be taken, and loads to be altered in the absence of an operator.

[1]Metallurgist and physicist, respectively, Fracture and Deformation Division, National Bureau of Standards, Boulder, CO 80303.

The automated test system minimizes testing time and operator attention. Data scatter is reduced owing to higher precision in crack length measurement and better control in data point spacing [4]. Because the testing is interactive and automatic in nature, the procedure is relatively easy to follow and requires minimal operator training. Finally, this approach eliminates subjective interpretation and influence of the experimenter.

The FCGR Test Method

The sequence of the FCGR test is: First, obtain the raw data, namely, fatigue crack length, a, versus elapsed fatigue cycles, N; then, reduce a and N data to a plot of da/dN versus ΔK, where da/dN is the FCGR in m/cycle and ΔK is the crack-tip stress-intensity factor range in MPa m$^{1/2}$. Typical outputs are presented in Fig. 1.

The number of elapsed fatigue cycles can be obtained from counters (electronic or mechanical) or conversion from time elapsed at the actual testing frequency. The methods of crack length measurement are complicated and have been a subject of extensive study [5,6]. Although several methods of crack length measurement have been developed, some require specialized equipment not commonly available in mechanical testing laboratories. The compliance technique, however, requires only monitoring of the load cell and the clip gage outputs, which is routinely achieved in mechanical testing. Compliance is defined as the specimen deflection per unit load, which is a function of crack length for a given material and specimen geometry. The load and deflection signals (voltages) can be interfaced to a computer. Because of the

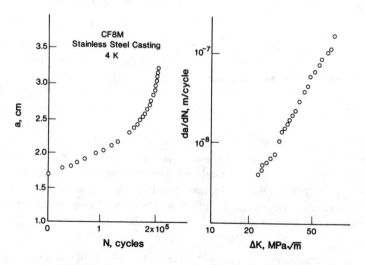

FIG. 1—*Data outputs from the automated FCGR test.*

simple instrumentation and the need in our laboratory for environmental chambers for cryogenic temperature and saltwater corrosion-fatigue tests, the compliance technique was chosen to measure the crack length.

Equipment for the Automated FCGR Test System

A schematic of the automated FCGR test system is shown in Fig. 2, which also shows the sequence of operation and interaction between various components. The test system consists of a closed-loop servo-controlled hydraulic mechanical testing machine, a programmable digital oscilloscope, a programmable arbitrary waveform generator, and a minicomputer.

The machine control unit, which is included in the hydraulic mechanical test machine, includes a servo-control system, a feedback system, two d-c conditioners, and a valve drive. A nonprogrammable function generator with an electronic pulse counter is usually built into the machine control unit of a commercially available mechanical testing machine. Signal amplifiers and a load cell are also included in the mechanical testing machine.

The programmable digital oscilloscope contains two 15-bit 100-kHz digitizers and it serves as an analog-to-digital (A/D) converter. In addition to its

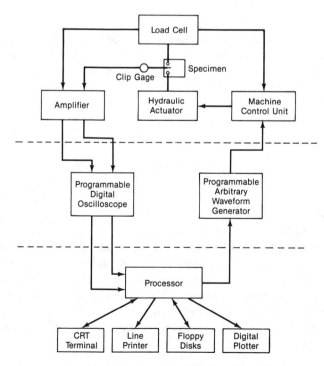

FIG. 2—*Schematic of the automated FCGR test system.*

high-speed A/D conversion rate, this oscilloscope features the ability to instantaneously freeze and hold data in memory. The problems encountered with slower A/D converters, such as interruptions during the test [7] and low test frequencies [8], are eliminated because of the high A/D conversion rate and the freeze-and-hold ability of this oscilloscope. For the near-threshold and the variable-load-amplitude FCGR tests, load levels vary with time and a programmable function generator is needed. For these tests, the programmable arbitrary waveform generator is used. The present programmable waveform generator is not connected to a cycle-counting device and the fatigue cycle counts are inferred from the cyclic frequency[2] and the time elapsed, as given by the computer. For the constant-load-amplitude FCGR test, the built-in function generator is used.

Included in the minicomputer are a cathode-ray-tube (CRT) terminal, a line printer, a dual floppy disk storage unit, and a digital plotter. The minicomputer uses the 16-bit word and has 128K words of memory. The minicomputer also contains an internal clock that reads to 1/60 s. The IEEE-488/ GPIB is used for the interface between the computer and the programmable digital oscilloscope and between the computer and the programmable arbitrary waveform generator.

Applications

In the following discussion, attention is focused on how the test system described above is used to conduct the FCGR tests. Requirements on grips, fixtures, specimen design, and specimen preparation are detailed in ASTM E 647-83 and other proposed standards [3,9] and are not discussed in this paper.

Constant-Load-Amplitude FCGR Test

The test system described in the previous section can be programmed to run the constant-load-amplitude FCGR test. The operational details, implementing the procedures set forth by ASTM E 647-83, are described in this subsection.

As shown in Fig. 3, the input parameters are fed by the operator into the computer through the CRT terminal. The input parameters include specimen identification, specimen dimensions, Young's modulus, selected time interval for measuring crack length, minimum load level for compliance measurement, load levels, and test frequency. The time interval for measuring crack length must be kept to a value small enough that every increment of crack growth will not exceed the recommended values as prescribed in ASTM E

---

[2]The frequency used for cycle calculations is checked with a frequency meter; the typical error in frequency is 50 ppm.

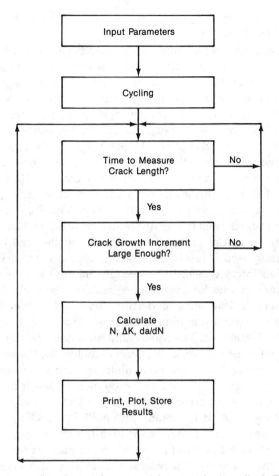

FIG. 3—*Summary flow chart of the automated constant-load-amplitude FCGR test.*

647-83. The minimum load level for compliance measurement is used to eliminate the possible crack closure effects [*10*], which have a significant effect on the accuracy of crack length measurement.

The precracked specimen is fatigue cycled under the prescribed loading conditions and cyclic frequency. A typical frequency is 10 Hz. When the preselected time interval (typical value is 1 min) for crack length measurement is reached, the computer requests the load-versus-deflection data from the programmable digital oscilloscope, which freezes the load-deflection data in the memory instantaneously, and correlates the data to a straight line using a linear least-squares fit. A linear correlation coefficient of 0.999 or better is usually obtained. From the resulting compliance, the instantaneous crack length is computed using the appropriate expression for the compliance calibration of the specimen [11]. The precision of the crack length measurement is typically within 0.04 mm.

The inferred crack length, which is obtained from the measured compliance, published compliance calibrations, and published Young's modulus, usually does not agree exactly with the actual crack length for a given material and specimen geometry. The exact reasons for the discrepancy between the inferred and the actual crack lengths are not clear and have been discussed in Refs 12 and 13. Correction factors to the compliance calibration have been used to obtain more accurate physical crack length predictions.

An alternative way of correcting the mismatch between the inferred and the actual crack lengths is to adopt an "effective modulus" for the material. The effective modulus, E_{eff}, is deduced from a known crack length in a given specimen geometry and compliance calibration. Typically, E_{eff} is deduced from one well-defined crack front that is visible on a post-test fracture surface. The crack front at the end of fatigue precracking or at the final fatigue crack length is generally used. The effective modulus approach, which is used in our laboratory, thus forces agreement between the inferred and the actual crack lengths and compensates for any errors regardless of source [13].

The portion of the load-versus-deflection curve used for compliance calculation is from the specified minimum load level to a value corresponding to 95% of the maximum load. The typical minimum load level used for compliance calculation is the mean load (load signal midpoint). It should be noted, however, that the specific value of the minimum load level used for a given material, specimen geometry, and load ratio must be larger than crack closure loads. The reason for excluding the upper 5% of the load for calculation is that the clip gage tends to vibrate, and noise in the clip-gage signal increases at the maximum load during the high-frequency test.

The increment of crack growth (the difference between the current measured crack length and the last recorded crack length) is checked against specified values which are within the recommended values of ASTM E 647-83. A value of 0.5 mm is typically specified for a 25.4-mm-thick standard compact-type specimen. If the increment of crack growth is equal to or greater than the specified value, the computer calculates N, ΔK, and da/dN; the digital plotter plots the data points (a, N) and $(da/dN, \Delta K)$ on the a-versus-N and on the da/dN-versus-ΔK graphs, such as shown in Fig. 1; the line printer prints the value of calculated compliance, the linear least-squares correlation coefficient, a, N, da/dN, and ΔK results. All the resulting data are stored on floppy disks for post-test analyses.

During the test, the point-to-point data reduction technique is used to calculate ΔK and da/dN from a and N. Usually the results are consistent with minimum scatter, as those shown in Fig. 1. If the results of ΔK versus da/dN scatter, the seven-point incremental polynomial method is used to smooth the results after the test is completed.

The computer programs for post-test analyses include the following capabilities:

1. reducing a-versus-N data to ΔK-versus-da/dN by the seven-point incremental polynomial method,
2. converting units,
3. plotting data in desired units,
4. plotting data in desired coordinate ranges,
5. plotting data for several different specimens on one graph (for comparison), and
6. calculating the material constants C and n in the Paris equation [14], $da/dN = C (\Delta K)^n$, and drawing the regression line through the data.

All computer programs, including the data acquisition routines, were written in the PDP-11 FORTRAN language.

Near-Threshold FCGR Test

The computer programs used in the constant-load-amplitude FCGR test, with some modifications, can be used for near-threshold FCGR tests. The major difference in procedures between the two tests is that the load levels in the near-threshold FCGR test decrease according to the initial ΔK-value (in the K-decreasing test technique). The load levels are calculated from the following equations [9]

$$\Delta K = \Delta K_0 \exp [C'(a - a_0)] \tag{1}$$

$$\Delta P = BW^{1/2} \Delta K / f_1(a/W) \quad \text{for compact-type specimen} \tag{2}$$

$$\Delta P = B \Delta K / f_2(a/W) \quad \text{for center-cracked-tension specimen} \tag{3}$$

$$P_{max} = \Delta P/(1 - R); P_{min} = P_{max} R \tag{4}$$

where

$$
\begin{aligned}
P_{max} &= \text{maximum load,} \\
P_{min} &= \text{minimum load,} \\
R &= P_{min}/P_{max}, \\
B &= \text{specimen thickness,} \\
W &= \text{specimen width,} \\
a &= \text{current crack length,} \\
a_0 &= \text{crack length at beginning of test,} \\
f_1(a/W) &= [2 + (a/W)][0.886 + 4.64(a/W) - 13.32(a/W)^2 + \\
&\quad 14.72(a/W)^3 - 5.6(a/W)^4]/[1 - (a/W)^{1.5}],
\end{aligned}
$$

$$f_2(a/W) = [(\pi a/W^2) \sec(\pi a/W)]^{1/2},$$
$$\Delta K = \text{current crack-tip stress-intensity range,}$$
$$\Delta K_0 = \text{crack-tip stress-intensity range at beginning of test, and}$$
$$C' = \text{negative constant.}$$

A typical value of C' is -0.08 mm^{-1}, which gives satisfactory results with no apparent anomalous crack growth for AISI 300-series stainless steels.

A flow chart describing the automated near-threshold FCGR test is summarized in Fig. 4. After each crack length measurement, the crack length is compared with the last stored crack length to ensure that a specified measur-

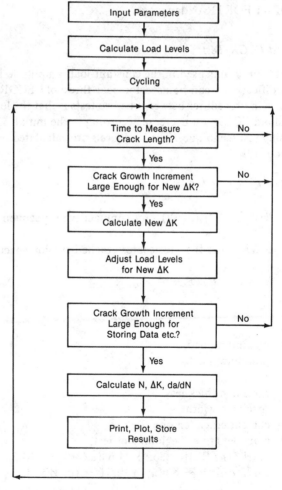

FIG. 4—*Summary flow chart of the automated near-threshold FCGR test.*

able small amount of crack growth has occurred. If this is not done, then some unnecessary load level adjustments will take place because of scatter in the crack length measurement. After the crack length has increased a certain amount (for example, 0.13 mm), the new ΔK is calculated according to Eq 1 and the new crack length is stored. The load levels are then adjusted using Eqs 2, 3, and 4.

In high-frequency fatigue testing, which is desirable in the near-threshold FCGR test, hydraulic lag might be a problem. This results in the specimen not being actually subjected to the load range commanded by the computer (or waveform generator). The problem is usually corrected by using proper signal conditioners and gain settings. However, overprogramming is sometimes necessary to overcome the persistent hydraulic lag. During the overprogramming process, which is done by trial-and-error method, the computer monitors the values of P_{max} and P_{min} through the programmable digital oscilloscope and makes necessary changes to achieve the desired values of P_{max} and P_{min}. The overprogramming is done whenever there is a hydraulic lag problem.

If the measured crack-growth increment, which is the difference between the current measured crack length and the last recorded crack length, is equal to or greater than specified values, which are within the recommended values [8], the values of N, ΔK, and da/dN are calculated and the results are printed, plotted, and stored. A value of 0.5 mm is typically specified for a 25.4-mm-thick standard compact-type specimen. The previously mentioned computer programs for post-test analyses are also applicable for analyzing the data obtained in the near-threshold FCGR test.

Variable-Load-Amplitude FCGR Test

The automated FCGR test system is also used in the variable-load-amplitude FCGR test. The procedures used in this application are similar to those described in the previous two subsections.

The computer reads the prerecorded load-time history from the floppy disks and controls the hydraulic machine through the programmable waveform generator. At a preselected time interval, the crack length is measured. A typical interval is 30 min for an average test frequency of 0.1 Hz. The desired output in the variable-load-amplitude FCGR test is time versus crack length. The results are printed, plotted, and stored for post-test analyses.

The present system has two limitations in the application of variable-load-amplitude FCGR testing. One is that the waveform generator needs about 0.1 s for changing one command to another, and this limits the average frequency to about 1 Hz. This will result in a situation of holding about 0.1 s at the peak loads when higher test frequencies are used. In a corrosive or a high-temperature environment, in which hold time at peak load is important, this might introduce anomalous fatigue crack growth. The other is the limited storage capacity of the floppy disks, which can store only a certain amount of load-time pairs. The present system uses soft disks, which can store about 18 000 load-time pairs. If longer load-time histories are desired, other means of storage devices such as hard disks must be used.

Summary

An automated FCGR test system has been developed that can be used for tests of constant-load-amplitude FCGR above 10^{-8} m/cycle (ASTM E 647-83), near-threshold FCGR, and variable-load-amplitude FCGR. The test system offers considerable time savings in data acquisition and in data reduction. The test procedure is relatively easy to follow and enables technicians to produce data with less scatter (with respect to the non-computer-aided technique), because higher precision in crack length measurement and better control in data point spacing are obtained, while manual data interpretation and data fitting are eliminated.

Acknowledgments

Mr. J. C. Moulder of the National Bureau of Standards is acknowledged for helpful discussions on the interface between the computer and the instruments. The work was supported by the Department of Interior, Minerals Management Service, and the Department of Energy, Office of Fusion Energy.

References

[1] Fatigue Thresholds: Fundamentals and Engineering Applications, J. Backlund, A. F. Blour, and C. J. Beevers, Eds., Engineering Materials Advisory Services, Chameleon Press, London, 1982.

[2] Bucci, R. J., "Development of a Proposed ASTM Standard Test Method for Near-Threshold Fatigue Crack Growth Rate Measurement," Fatigue Crack Growth Measurement and Data Analysis, ASTM STP 738, S. J. Hudak, Jr., and R. J. Bucci, Eds., American Society for Testing and Materials, Philadelphia, 1981, pp. 5–28.

[3] Crooker, T. W., Bogar, F. D., and Yoder, G. R., "Standard Method of Test for Constant-Load-Amplitude Fatigue Crack Growth Rates in Marine Environments," NRL Memorandum Report 4594, Research Laboratory, Washington, DC, Aug. 6, 1981.

[4] Wei, R. P., Wei, W., and Miller, G. A., "Effect of Measurement Precision and Data-Processing Procedure on Variability in Fatigue Crack Growth-Rate Data," Journal of Testing and Evaluation, Vol. 7, No. 2, March 1979, pp. 90–95.

[5] The Measurement of Crack Length and Shape During Fracture and Fatigue, C. J. Beevers, Ed., Engineering Materials Advisory Services, Chameleon Press, London, 1980.

[6] Advances in Crack Length Measurement, C. J. Beevers, Ed., Engineering Materials Advisory Services, Chameleon Press, London, 1981.

[7] Cheng, Y.-W., "A Computer-Interactive Fatigue Crack Growth Rate Test Procedure," Materials Studies for Magnetic Fusion Energy Applications at Low Temperatures—VI, R. P. Reed and N. J. Simon, Eds., NBSIR 83-1690, National Bureau of Standards, Boulder, CO, 1983, pp. 41–51.

[8] Ruschau, J. J., "Fatigue Crack Growth Rate Data Acquisition System for Linear and Non-

linear Fracture Mechanics Applications," *Journal of Testing and Evaluation*, Vol. 9, No. 6, Nov. 1981, pp. 317–323.

[9] "Proposed ASTM Test Method for Measurement of Fatigue Crack Growth Rates," *Fatigue Crack Growth Measurement and Data Analysis, ASTM STP 738*, S. J. Hudak, Jr. and R. J. Bucci, Eds., American Society for Testing and Materials, Philadelphia, 1981, pp. 340–356.

[10] Elber, W., "The Significance of Fatigue Crack Closure," *Damage Tolerance in Aircraft Structures, ASTM STP 486*, M. S. Rosenfeld, Ed., American Society for Testing and Materials, Philadelphia, 1971, pp. 230–242.

[11] Hudak, S. J., Jr., Saxena, A., Bucci, R. J., and Malcolm, R. C., "Development of Standards of Testing and Analyzing Fatigue Crack Growth Rate Data," AFML-TR-78-40, Air Force Materials Laboratory, Wright-Patterson Air Force Base, Ohio, May 1978.

[12] Nicholas, T., Ashbaugh, N. E., and Weerasooriya, T., "On the Use of Compliance for Determining Crack Length in the Inelastic Range," *Fracture Mechanics: Fifteenth Symposium, ASTM STP 833*, R. J. Sanford, Ed., American Society for Testing and Materials, Philadelphia, 1984, pp. 682–698.

[13] Tobler, R. L., and Carpenter, W. C., "A Numerical and Experimental Verification of Compliance Functions for Compact Specimens," to be published in *Engineering Fracture Mechanics*.

[14] Paris, P. C., and Erdogan, F., "A Critical Analysis of Crack Propagation Laws," *Transactions*, American Society of Mechanical Engineers, *Journal of Basic Engineering*, Series D, Vol. 85, No. 3, 1963, pp. 528–534.

Systems for Fracture Testing

James A. Joyce[1] and Gerald E. Sutton[2]

An Automated Method of Computer-Controlled Low-Cycle Fatigue Crack Growth Testing Using the Elastic-Plastic Parameter Cyclic *J*

REFERENCE: Joyce, J. A. and Sutton, G. E., "**An Automated Method of Computer-Controlled Low-Cycle Fatigue Crack Growth Testing Using the Elastic-Plastic Parameter Cyclic *J*,**" *Automated Test Methods for Fracture and Fatigue Crack Growth, ASTM STP 877*, W. H. Cullen, R. W. Landgraf, L. R. Kaisand, and J. H. Underwood, Eds., American Society for Testing and Materials, 1985, pp. 227–247.

ABSTRACT: The application of J-integral methods to fatigue crack growth testing was first presented by Dowling and Begley [*1*]. They demonstrated that under certain conditions the J-integral range (ΔJ) applied to a cracked body correlates well with the measured crack growth per cycle in tests performed in the elastic-plastic regime. This paper presents an automated test system for computer-controlled low-cycle fatigue crack growth evaluation of ductile materials using cyclic-*J*, and details the procedure for this technique. The system consists of a hydraulic test machine, a minicomputer, and a microcomputer. The test machine is controlled in either clipgage or stroke mode with the control signal of the minicomputer channeled through a digital-to-analog converter. The microcomputer is dedicated to compliance calculation. Analysis of crack length, closure load, and applied ΔJ is performed each cycle and the control signal is adjusted to achieve the desired *J*-range. Crack growth results are periodically stored on disk along with load-crack opening displacement data for the cycle.

This automated system has been utilized for rising *J*-range, *J*-range shedding, and constant *J*-range tests using compact tension specimens of two high-strength structural steels. Results from these tests show agreement with the linear extrapolation of crack growth data obtained under linear-elastic conditions, and demonstrate the promise of such an approach for low-cycle fatigue crack growth characterization.

KEY WORDS: low-cycle fatigue crack growth, cyclic *J*, J-integral range, elastic-plastic fracture, automated testing, crack propagation

[1]Associate Professor, Mechanical Engineering Department, U.S. Naval Academy, Annapolis, MD 21402.
[2]Metallurgist, David W. Taylor Naval Ship Research and Development Center, Bethesda, MD 20084.

The application of J-integral methods to fatigue crack growth rate testing was first presented by Dowling and Begley [1]. They demonstrated that under certain conditions, the J-integral range applied to a cracked body correlates positively with the measured crack growth per cycle in tests performed in the elastic-plastic regime. Tanaka and co-workers [2] used the J-integral range to study the effect of loading condition and specimen geometry on elastic-plastic fatigue crack growth in low-carbon steel. Using thin specimens and applying high cyclic stresses they reported a significant acceleration in crack growth which deviated from the stable relation between growth rate and J-range. This acceleration coincided with the spread of plastic yielding across the ligament of the specimen. They found that the beginning of crack growth acceleration at high cyclic stresses could be predicted by correlating maximum load to limit load. El Haddad and Mukherjee [3] performed cyclic J-range tests in load control and in deflection control. Controlled at constant load amplitude, the crack growth rate increased, while for constant displacement amplitude the crack growth rate decreased. The displacement-controlled tests produced macroscopic crack closure, and the J-range was determined from the area under the load-displacement curve, above the crack opening load. El Haddad and Mukherjee attributed their success in correlating increasing crack growth rates with decreasing crack growth rates to proper accounting for crack closure in the J-integral analysis. Although questions remain regarding the applicability of the J-integral to the characterization of the fatigue crack growth process [4], this fracture parameter provides valuable information on the intensity of the stress and strain fields near the crack tip of a crack loaded in tension, and if cycling effects do not become dominant, these fields should control the crack growth resulting from individual cycles. Other advantages of a J-integral based methodology are that this parameter can be easily measured in the laboratory for simple specimen geometries, and can be evaluated for flawed structures and structural elements.

The objective of this paper is to describe an automated test system for computer-controlled low-cycle fatigue crack growth using the J-integral, and to present preliminary test results. In this system, the computer is used to control the applied J-range so that conditions of increasing J-range, constant J-range, and decreasing J-range can be applied to the test specimen as desired. This system shows promise of being a powerful tool for evaluation of crack growth rates in the elastic-plastic regime, but it will also be useful in determining specimen size limitations and history/sequence effects on elastic-plastic fatigue crack growth rate.

Test Procedure

Materials

Two nickel alloy steels have been tested to date, HY-80 and HY-130. Plate supplied in 50-mm (2 in.) thickness was used for all tests. The chemical compositions and mechanical properties of the plates are presented in Tables 1 and 2, respectively.

J-Integral Analysis

The specimen geometry used exclusively in this work has been a compact specimen for which the J-integral can be calculated using the analysis of Merkle and Corten [6] from the expression

$$J = \frac{\beta A}{Bb} \qquad (1)$$

where

a = crack length,
B = specimen thickness,
b = uncracked ligament dimension,
A = area under load (load point displacement curve),

and β is obtained from a modified form of the Merkle-Corten analysis by Clarke and Landes [5] as

$$\beta = 2[(1 + \alpha)/(1 + \alpha^2)] \qquad (2)$$

with

$$\alpha = \{[2(a/b)]^2 + 2[2(a/b)] + 2\}^{1/2} - [2(a/b) + 1] \qquad (3)$$

The area under the load displacement curve used for J evaluation is that present above the crack closure load as depicted by the hatched region in the schematic of Fig. 1. An accurate determination of this closure load is a very important aspect of determining the J-range applicable during a fatigue cy-

TABLE 1—*Chemical compositions (weight %)*.

Material	C	Mn	P	S	Si	Ni	Cr	Mo	V	Al
HY-80	0.15	0.33	0.010	0.014	0.21	2.10	1.25	0.24	0.003	...
HY-130	0.12	0.78	0.005	0.005	0.30	4.97	0.58	0.59	0.06	0.04

TABLE 2—*Mechanical properties*.

Material	0.2% Yield Strength, MPa (ksi)	Ultimate Tensile Strength, MPa (ksi)	Elongation, % in 50.8 mm (2 in.)	Reduction of Area, %
HY-80	607 (88)	717 (104)	20	65
HY-130	944 (137)	1006 (146)	19	63

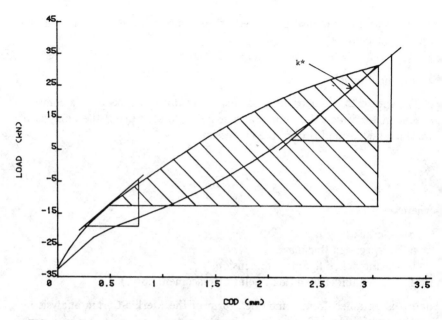

FIG. 1—*Schematic showing area of load-COD record used to calculate the applied J-range and the slope method for determining closure load.*

cle. The technique used in this work is to locate the point on the loading side of the cycle at which the slope of the load displacement curve falls to the value of the slope found over the first region of unloading of the previous cycle, labelled k^* in Fig. 1. It is demonstrated in the discussion to be presented later that this closure load can vary from minimum load to within 20% of the maximum load depending on the applied J-range, whether the applied J-range is increasing or decreasing, and on the crack length.

The basic concept of J-integral in fatigue crack growth is that the applied J-range during a cycle of loading will produce a repeatable amount of fatigue crack growth in a given material—and different amounts of crack growth in different materials. Following work done in the elastic regime, crack growth rate is plotted against the applied J-range to give a relationship similar to the Paris law, implying a correlation of the form

$$\frac{da}{dN} = A(\Delta J)^n \tag{4}$$

Data of this type was obtained for ASTM A533B steel by Dowling and Begley [1] in their early work. Other investigators [7-10] have applied this empirical relationship to model crack propagation in a variety of ductile materials. Many questions were left unanswered in the early work by Dowling and Begley. Principal among them were whether crack growth rate data in the elastic-plastic region was really geometry independent, whether it was dependent on the average J-level, what the limits on J might be for a particular geometry, and the effects of nonsinusoidal loading. All of these questions must be answered before serious attempts can be used to apply the J-integral method to practical engineering problems.

The major thrust of this work is to study the geometry dependence of the J-integral, but before even this can be undertaken, an experimental methodology for evaluating data under J-integral control was required and this methodology is the subject of the remainder of this report.

Experimental Description

The specimens used in these tests were compact specimens with a 2T plan configuration as shown in Fig. 2. As is common in J-integral testing, the specimen was notched to allow the measurement of crack opening displacement (COD) on the load line. Razor blade knife edges were fixed on the load line on all specimens to reduce the hysteresis and to allow accurate COD, compli-

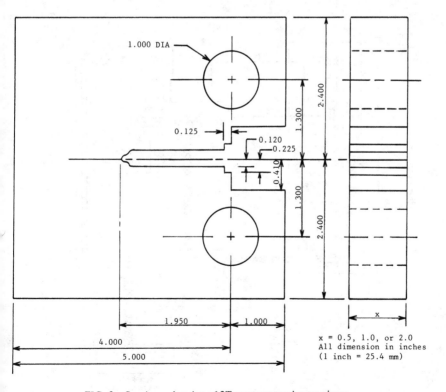

FIG. 2—*Specimen drawing of 2T compact tension specimen.*

ance, and hence crack length measurements. Three specimen thicknesses were available—13, 26, and 52 mm ($\frac{1}{2}$, 1, and 2 in.)—so that an initial study of specimen size effects could be made. All specimens were fabricated in the transverse-longitudinal (T-L) orientation. Fatigue crack growth tests were conducted in a computer-controlled hydraulic test apparatus, shown schematically in Fig. 3. Two separate test machines have in fact been utilized in this investigation. One is a standard MTS Model 810 Automated Test System utilizing a PDP 11/34A minicomputer and BASIC language software. The system has a 100-kN capacity. The second system includes a 1330 kN Wiedemann hydraulic test machine controlled via a Tektronix 4052 microprocessor system, also using the BASIC language. Because of the limited processor speed of the second system, an additional Z80 microprocessor is utilized to measure the specimen compliance over a specified load range and to transfer this value to the 4052 controller when requested. The schematic of Fig. 3 shows the two-processor system.

All fatigue cycles were of the displacement-control type, with the test machine returning to an initial COD setting as the zero point at the start of the test. The MTS system controls on the COD signal while the Wiedemann sys-

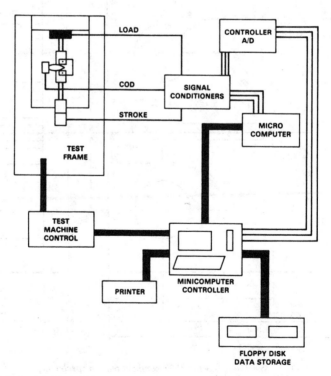

FIG. 3—*Schematic of automated low-cycle fatigue crack growth rate test system.*

tem controls on the stroke signal. During an individual cycle, the minicomputer reads load-COD data at fixed time intervals for the complete cycle, with a cycle taking between 20 and 40 s to complete. Load-COD data, involving approximately 45 data pairs, are taken over the first portion of the load cycle for subsequent calculation of a crack closure load. High-density data involving more than 2000 data pairs are taken by the microprocessor over the first portion of the unloading for subsequent calculation of the specimen crack length. The upper portion of the hysteresis curve (the loading side) and the lower portion (the unloading side) are scanned with only enough points—approximately 27 in each case—to give accurate measurement of the area needed for J calculation. At the completion of each cycle, the computer calculates the present crack length, determines the crack closure load, and then calculates J from the area under the load-COD curve but above the closure load line as shown in Fig. 1. At this point, control parameters are changed if necessary to hold the test near a desired target J-range. Data are stored on magnetic disk if crack growth has occurred. The test can be stopped if shutoff values have been exceeded, or another cycle can be initiated. By varying the control parameters, specifically COD or stroke as applicable, the system can obtain conditions of slowly increasing ΔJ, constant ΔJ, decreasing ΔJ, or stepped ΔJ.

In the start-up program menu, the operator designates the increment by which the J-range will be changed for an increment of crack extension. For the two materials in this paper the J-range increments were approximately 2.63 kPa·m (15 in.·lb/in.^{2}) and this increment of change was applied to the target value each time there was measured crack growth of 0.1 mm (0.004 in.).

The specimen crack length is obtained from the compliance measured over the first 30% to 40% of load drop during each cycle. This lower limit on data used for crack length estimation can be raised somewhat by the program if the crack closure load is found to move close to this level. The crack closure point is found by fitting a quadratic polynomial to a "window" of uploading data, and then calculating the load and COD value within the window at which the specimen loading compliance equals the compliance measured (as described above) over the first 30% to 40% of unloading. For the first cycle the window includes the minimum load; thereafter it adjusts automatically to keep the detected crack closure point near the window center. If the load-COD record is very nearly elastic with no discernible curvature, the program designates the closure point at the minimum load measured.

Both Dowling and Begley [1] and Mowbray [11] defined the crack closure load using an inflection point obtained on the unloading side of the load-COD hysteresis loop. The inflection was found by inspection and was, therefore, very subjective. Application of a similar method was used in some cases here in which the computer fit a cubic polynomial to the unloading side of the load-COD hysteresis loop and identified the inflection point as the point where the curvature of the cubic polynomial was equal to zero.

Test data are stored on computer disk approximately every fourth cycle when crack growth is slow, and the storage frequency increases to every cycle when crack growth is rapid. This includes load-COD data pairs for the complete fatigue cycle, as well as the calculated compliance value, crack length, closure load, ΔJ, and the cycle number. These data are subsequently analyzed using an iterated polynomial curve-fitting and differentiation scheme similar to that used in linear-elastic fatigue data reduction [12] which calculates the crack growth per cycle as a function of the cycle count, ΔJ, crack length, or whatever, producing an output file for computer graphics display of the final results. A series of results obtained by this process is presented in the following section.

Discussion of Results

This automated test system has been developed as a research tool to allow determination of the applicability of the J-integral to elastic-plastic fatigue crack growth. For this reason, many parameters can be varied. For example, the type of test control can be selected, including increasing ΔJ, constant ΔJ, decreasing ΔJ, stepped ΔJ, and simple spectra of applied ΔJ. Other quantities which can be varied include the rate of ΔJ adjustment and the rate of cyclic loading.

Uniformly Varying ΔJ

In the first application of this test to materials characterization, crack growth rate tests were conducted with slowly varying ΔJ conditions, sweeping from elastic to nearly fully plastic loading conditions. Results of two such tests on the materials used in this study are shown in Figs. 4 and 5. Both ΔJ increasing and ΔJ decreasing data are displayed in these figures, along with baseline fatigue crack growth rates obtained under linear-elastic conditions using high cycle fatigue crack growth rate test procedures. Good agreement exists between the linear-elastic high-cycle data and the ΔJ decreasing results. However, there is a distinct deviation of the ΔJ decreasing crack growth rate from the ΔJ increasing. This is believed to result from inaccurate calculation of the applied J-integral range for the condition of ΔJ increasing in which the load-COD record was so nearly straight that no slope (or inflection point) defined crack closure point was determinable. In this case the closure load was taken as the minimum load for the cycle, thereby overpredicting the applied J-range. Therefore, the actual applied J-range for ΔJ increasing tests of Figs. 4 and 5 is less than that shown.

To demonstrate this point further, the same specimens tested in Figs. 4 and 5 were tested a second time under ΔJ increasing conditions. The results are presented in Figs. 6 and 7. This time a large plastic zone had to be overcome on each cycle to return to the initial displacement setting. The compressive

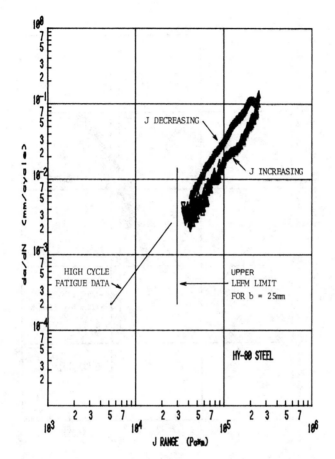

FIG. 4—*Crack growth rate data for HY-80 steel using the automated test system with* ΔJ *increasing followed by* ΔJ *decreasing.*

loads needed to overcome the plasticity at the crack tip produced the "knee" in the unloading side of the hysteresis loop, as shown in Fig. 1. As a result, distinct closure loads were identified for each cycle and, after an initial retardation associated with low initial J-range, the crack growth rates for the subsequent ΔJ increasing tests of Figs. 6 and 7 correspond well with the ΔJ decreasing results of Figs. 4 and 5, respectively. This emphasizes the importance of a distinctly definable closure load for the accurate determination of the applied J-integral range.

Hysteresis loops demonstrating distinct and indistinct crack closure loads are shown in Figs. 8a–8c. For the initial ΔJ increasing case shown in Fig. 8a the load-COD hysteresis loops are very symmetrical and show no "knee" which can be related to crack closure. Only when ΔJ exceeded 227 kPa·m (1300 in.-lb/in.²), corresponding to crack growth rates of 0.05 mm/cycle, did

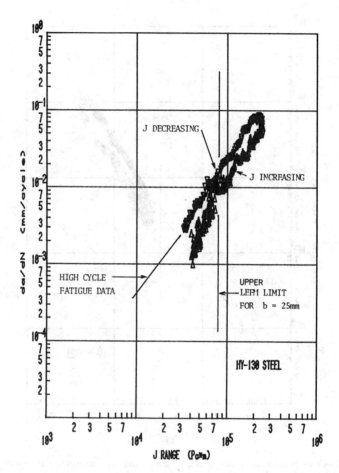

FIG. 5—*Crack growth rate data for HY-130 steel using the automated test system with* ΔJ *increasing followed by* ΔJ *decreasing.*

the program detect a closure load by the slope method described above. The ΔJ quantity used for this test by the computer was based on minimum load, but a more correct closure load appears to exist between the minimum load and zero load. For the two tests with load-COD hysteresis loops shown in Figs. 8b and 8c crack closure loads were always available to the computer and for these tests good agreement in crack growth rate exists between the ΔJ decreasing and the ΔJ increasing tests.

Even further support of this point is demonstrated by a reanalysis of the initial ΔJ increasing data shown in Fig. 5 using zero load as the crack closure load for ΔJ calculation. The result is shown in Fig. 9 where the original crack growth rate data and the new calculation based on zero crack closure load bracket the ΔJ decreasing results of Fig. 4. This is thought here to demon-

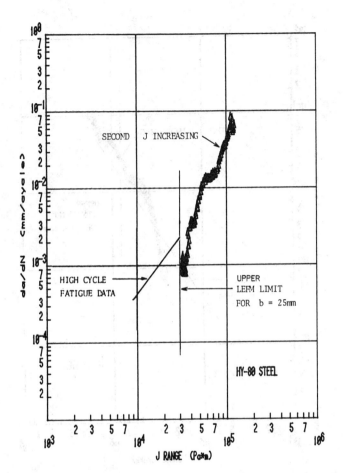

FIG. 6—*Crack growth rate data for HY-80 steel for the specimen of Fig. 4 subjected to a second* ΔJ *increasing test.*

strate that a crack closure load does exist for the initial ΔJ increasing test at a value between the minimum load encountered on a given cycle and zero load. The two choices of crack closure point, zero load and minimum load, used for ΔJ increasing tests bound the real crack closure point and give *da/dN* versus ΔJ results on opposite sides of the ΔJ decreasing results. This closure load does not, however, appear to be obtainable from load and COD data alone.

Because the closure load was not defined by the slope method described above during the early cycles of ΔJ increasing tests, the inflection point method was tried. Results showed that the inflection point method also failed to define closure loads when nearly symmetrical load-COD hysteresis loops were present, which is the case during ΔJ increasing tests. The slope method proved to be more sensitive in defining the crack closure load than the inflec-

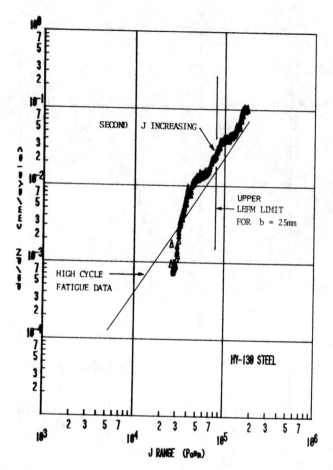

FIG. 7—*Crack growth rate data for HY-130 steel for the specimen of Fig. 5 subjected to a second* ΔJ *increasing test.*

tion point method in that it consistently picked up a closure load other than minimum load as soon as small amounts of nonsymmetry were present. Once both methods identified closure loads, excellent agreement existed between the results of the two methods.

Constant ΔJ

The crack length versus cycle count output data for a specimen controlled at a constant $\Delta J = 92.8 \, \text{kPa} \cdot \text{m}$ (530 in. $\cdot$ lb/in.²) are shown in Fig. 10. Ideally one would expect a straight line here, but what is found is a piecewise linear plot with three distinct slopes. This result is a further demonstration of the

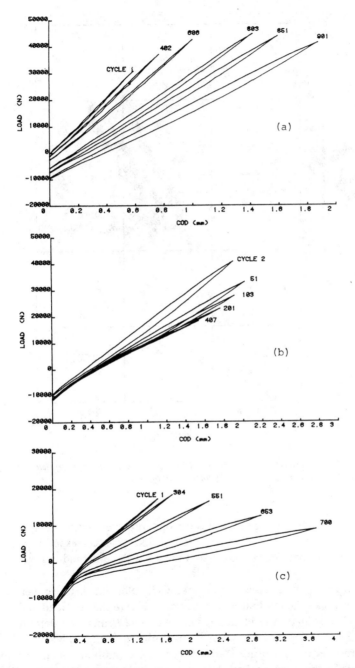

FIG. 8—*Load versus load-point displacement for a compact specimen of HY-130 steel for which (a) J-range was uniformly increased, (b) then uniformly decreased, then (c) subsequently increased again.*

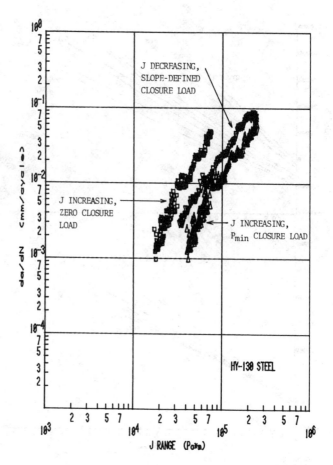

FIG. 9—*Crack growth rate data for HY-130 steel showing that slope-defined closure load is between P_{min} closure load and a crack closure load of zero.*

difficulty in defining the crack closure load discussed in the previous section. The distinct change in slope shown at $N \cong 730$ cycles corresponds to the cycle at which the load-COD records developed a slope-definable crack closure value.

Figure 11 shows a series of cycle load-COD plots for this specimen. For the first few cycles the minimum load was very close to the zero load and thus the crack closure load was well defined and a correct J-range was determined and used. By about 50 cycles, however, the minimum load had become distinctly compressive, but a closure load other than the minimum load was not discernible by the slope method. Using successively greater compressive closure loads caused the area calculation for the cycle to increase, and this in turn resulted in increasing calculated values of ΔJ. To adjust to the target ΔJ, the

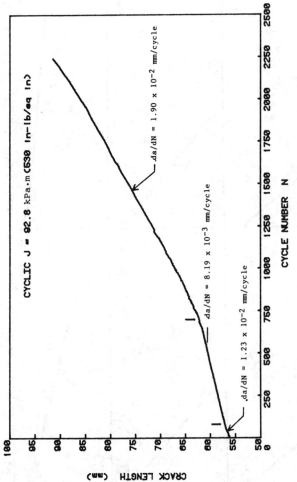

FIG. 10—*Record of crack length versus cycles for HY-130 steel at a constant applied J-range.*

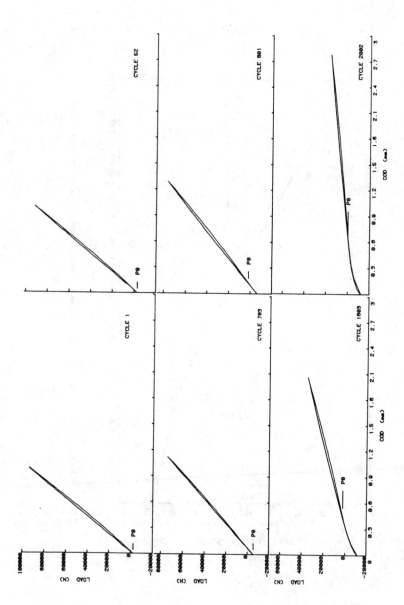

FIG. 11—*Load versus load-line displacement for cycles of a constant* ΔJ *test* ($\Delta J = 92.8 \, kPa \cdot m$ *(530 in.* $\cdot lb/in.$) *which shows the change in closure load from the minimum load to some slope-defined closure load.*

computer reduced the control COD range, producing the lower crack growth rate seen in Fig. 10, which is the result of the lower applied *J*-range. As cycling continued, crack-tip plasticity increased, and the load-COD loops slowly developed a larger hysteresis loop and a slope-defined crack closure point became measurable (Fig. 11). The program then automatically adjusted the COD to larger values since the higher closure load resulted in calculated *J*-range values lower than the target *J*-range. A higher crack growth rate resulted from increasing the applied *J*-range to the target *J*-range, and from approximately cycle 730 onward the correct closure point and corresponding correct *J*-range were always used. This produced a very constant crack growth rate as the crack grew across the major part of the specimen, that is, from $a/W = 0.6$ to $a/W = 0.9$.

Stepped J

Figure 12 shows crack length plotted versus cycle number for a test in which ΔJ was held constant for set amounts of crack extension and then increased stepwise three times by a relatively large amount, and then reduced stepwise in the same size steps. The step size was on the order of 50% of the initial *J*-range of 52 kPa·m (300 in.·lb/in.[2]). Tic marks show the cycle numbers where the target ΔJ was changed; however, it took between 10 and 50 cycles beyond this before the applied ΔJ range reached the new target value. The crack growth rates were independently determined (using a least-squares routine) for each segment of the *a–N* curve of Fig. 12. The growth rates corresponding to each step in the test are shown on the *a–N* trace.

Figure 13 presents the crack growth rate data for the constant *J*-range test of Fig. 10 and the stepped *J*-range test of Fig. 12. The baseline data are represented by the dashed line and are taken from the high-cycle linear-elastic data combined with the elastic-plastic data from the *J*-range decreasing test of Fig. 5. The constant *J* test of Fig. 10 produced two distinct crack growth rates, one where the applied *J*-range was overpredicted, and a higher growth rate once closure load and applied *J*-range were accurately evaluated. In Fig. 13 these growth rates are consistent with *J* increasing and *J* shedding data, respectively. Also shown in Fig. 13 are the crack growth rate results of the stepped *J*-range test of Fig. 12. The first four *J*-range increasing steps are seen to correspond to the previous *J*-range increasing data while the next three *J*-range decreasing steps show a transition back toward the *J*-range baseline data. These results are consistent with the previous observations that for ΔJ increasing tests accurate closure loads cannot be obtained by slope or inflection techniques from the load-COD record alone and that use of minimum load in these cases for the ΔJ calculation leads to an overestimate of the true applied ΔJ. As plasticity develops or as the *J*-range is shed, accurate closure loads become measurable and the measured crack growth rate results again become consistent with baseline data.

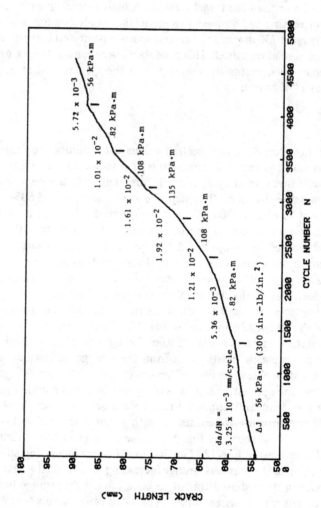

FIG. 12—*Record of crack length versus cycles for HY-130 steel showing crack growth rate response in a test in which J was held constant over set amounts of crack extension, then suddenly stepped up/down as indicated.*

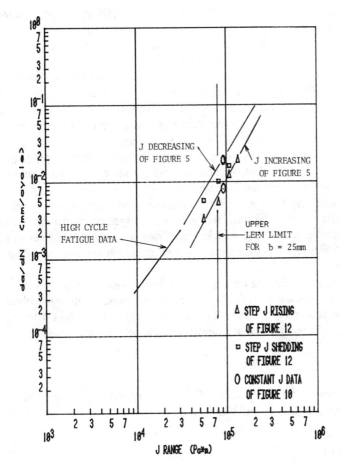

FIG. 13—*Crack growth rate versus J-range for HY-130 steel comparing stepped J-range results and constant J-range results with the baseline data.*

Summary

The principal objective of this paper was to describe an automated test system for computer-controlled low-cycle fatigue crack growth, and present results which show the degree of success achieved with this system. This automated system calculates the J-integral range in real time and adjusts the test machine control quantity (stroke or COD) to produce test conditions close to the *J*-range desired. With this automated test method the target *J*-range can then be held constant or varied in a number of ways to allow the researcher to investigate a wide range of effects.

The test system described herein was used to produce results for two types of steel, and the agreement with growth rate data in the linear elastic range was demonstrated. During testing and data analysis the following key observations were made:

1. Calculation of the correct applied J-integral range requires accurate determination of the crack closure load.

2. Crack closure load can be identified for the load-COD hysteresis loop using either the slope method or an inflection point method in most instances. However, for ΔJ increasing tests where plasticity at the crack tip is minimal or nonexistent and the corresponding hysteresis loop is very symmetrical, neither method clearly identifies the correct closure load. In these instances the use of the minimum load for calculation of J-range introduces appreciable error with the result that the actual applied J-integral range is substantially lower than the calculated J-range.

3. Decreasing J-range tests are consistent with linear-elastic crack growth rate results obtained by standard high-cycle fatigue crack growth rate testing. This is attributed to the success in identifying the closure load once significant plastic deformation is overcome at the crack tip during the unloading side of each cycle.

4. J-range increasing tests (performed with specimens previously tested in a J-range decreasing mode) for which crack closure loads were obtained by the slope procedure gave crack growth rate results consistent with high-cycle linear-elastic results and elastic-plastic J-range decreasing results.

Acknowledgments

The authors gratefully acknowledge the support of the Naval Sea Systems Command and Dr. H. H. Vanderveldt.

References

[1] Dowling, N. E. and Begley, J. A., "Fatigue Crack Growth During Gross Plasticity and the J-Integral," *Mechanics of Crack Growth, ASTM STP 590*, American Society for Testing and Materials, Philadelphia, 1976, pp. 82–103.

[2] Tanaka, T., Hoshide, T., and Nakata, M., "Elastic-Plastic Crack Propagation Under High Cyclic Stresses," *Elastic-Plastic Fracture: Second Symposium, ASTM STP 803*, C. F. Shih and J. P. Gudas, Eds., American Society for Testing and Materials, Philadelphia, 1983, pp. II-708–II-722.

[3] El Haddad, M. H. and Mukherjee, B., "Elastic-Plastic Fracture Mechanics Analysis of Fatigue Crack Growth," *Elastic-Plastic Fracture: Second Symposium, ASTM STP 803*, C. F. Shih and J. P. Gudas, Eds., American Society for Testing and Materials, Philadelphia, 1983, pp. II-689–II-707.

[4] Leis, B. N. and Zahoor, A., "Cyclic Inelastic Deformation Aspects of Fatigue Crack Growth Analysis," *Fracture Mechanics: Twelfth Conference, ASTM STP 700*, American Society for Testing and Materials, Philadelphia, 1980, pp. 65–96.

[5] Merkle, J. G. and Corten, H. T., "A J-Integral Analysis for the Compact Specimen Considering Axial Force as Well as Bending Effects," *Journal of Pressure Vessel Technology, Transactions*, American Society of Mechanical Engineers, Nov. 1974, pp. 286–292.

[6] Clarke, G. A. and Landes, J. D., "Evaluation of J for the Compact Specimen," *Journal of Testing and Evaluation*, Vol. 7, No. 5, Sept. 1979, pp. 264–269.

[7] Huang, J. S. and Pelloux, R. M., "Low Cycle Fatigue Crack Propagation in Hastelloy-X at 25 and 700C," *Metallurgical Transactions A*, Vol. 11, June 1980, pp. 899-904.

[8] Taira, S., Tanaka, K., and Hoshide, T., "Evaluation of J-Integral Range and Its Relation to Fatigue Crack Growth Rate," 22nd Japan Congress on Materials Research, Kyoto, Japan, Sept. 1978; also, *Proceedings*, Society of Materials Science, 1979, pp. 123-129.

[9] Kaisand, L. R. and Mowbray, D. F., "Relationships Between Low-Cycle Fatigue and Fatigue Crack Growth Rate Properties," *Journal of Testing and Evaluation*, Vol. 7, No. 5, Sept. 1979, pp. 270-280.

[10] Starkey, M. S., and Skelton, R. P., "A Comparison of the Strain Intensity and Cyclic J Approaches to Crack Growth," *Fatigue of Engineering Materials and Structures*, Vol. 5, No. 4, 1982, pp. 329-341.

[11] Mowbray, D. F., "Use of a Compact-Type Strip Specimen for Fatigue Crack Growth Rate Testing in the High-Rate Regime," *Elastic-Plastic Fracture, ASTM STP 668*, J. D. Landes, J. A. Begley, and G. A. Clarke, Eds., American Society for Testing and Materials, Philadelphia, 1979, pp. 736-752.

[12] Hudak, S. J., Jr., Saxena, A., Bucci, R. J., and Malcolm, R. L., "Development of Standard Methods of Testing and Analyzing Fatigue Crack Growth Rate Data," Technical Report AFM-TR-78-40, Air Force Materials Laboratory, Wright-Patterson Air Force Base, Ohio, May 1978.

Mitchell Jolles[1]

Automated Technique for R-Curve Testing and Analysis

REFERENCE: Jolles, M., "**Automated Technique for R-Curve Testing and Analysis,**" *Automated Test Methods for Fracture and Fatigue Crack Growth, ASTM STP 877,* W. H. Cullen, R. W. Landgraf, L. R. Kaisand, and J. H. Underwood, Eds., American Society for Testing and Materials, Philadelphia, 1985, pp. 248–259.

ABSTRACT: The crack growth resistance curve (R-curve) appears to be a viable approach to elastic-plastic fracture analysis. An automated technique for R-curve testing and analysis provides the opportunity to enhance accuracy and resolution while reducing variability due to influences of the test operator and data analyst. A microcomputer-based test control, data acquisition, and analysis system is described. Details of the test hardware, procedures utilized for the conduct of R-curve experiments, and methods for the analysis of the data are presented. The results demonstrate the quality and reproducibility obtained by the technique.

KEY WORDS: fracture mechanics, elastic-plastic fracture, R-curve, data acquisition

Requirements for the development of a practical methodology for the prediction of crack growth initiation, stable extension, and instability include the accurate characterization of material resistance to fracture under nominally inelastic conditions. This requires the determination of an appropriate elastic-plastic fracture parameter [such as J or crack-tip opening displacement (CTOD)] as a function of stable flaw growth. Such an R-curve concept appears to be a viable approach to ductile fracture analysis.

R-curve testing and analysis has received great attention, and results abound in the literature. Observation of the results indicate that in many instances considerable scatter occurs both in the data within a single R-curve and in test replication. Accurate determination of the initiation toughness is often difficult. This is primarily due to the test procedure utilized, such as the uncertainty associated with crack length measurements using unloading compliance, and the method of data acquisition and analysis.

---

[1]Head, Fracture Mechanics Section, Naval Research Laboratory, Washington, DC 20375.

An automated technique for R-curve testing and analysis which overcomes these difficulties has been developed. In addition, the procedures allow the simultaneous determination of a number of fracture parameters and R-curve presentations from a single test record. The technique uses hardware and test procedures to maximize accuracy and resolution while the automation ensures minimization of variability due to influences of the test operator and data analyst. The technique utilizes a microcomputer-based test control, data acquisition and analysis system which includes an eight-inch (20 cm) floppy disk for data storage, cathode ray tube (CRT) and plotter graphics, relay actuator for test equipment activation control, and 16-bit digital voltmeter and 12-bit analog-to-digital conversion for data acquisition. The test procedures include the monitoring of load measured by a standard load cell, crack surface displacement measured by a capacitance displacement transducer, and direct-current electric potential measurement for quantifying crack extension and determining the point of crack growth initiation. In the sequel, the hardware, test procedure, and method of data analysis are described and examples of results presented.

Computational Considerations

In order to complement and provide background appropriate for later discussions of procedure and hardware selection, a review of computations used to establish an R-curve is appropriate.

Crack Extension

Direct-current electric potential is used to quantify crack extension. As shown in Fig. 1, the current is introduced at the midwidth of the specimen and the potential measured at the front face just above and below the notch. This is the optimum configuration after considering accuracy, sensitivity, and reproducibility [1], and allows simply computation of the crack length from the measured potential, V by [2]

$$a = \frac{2W}{\pi} \cos^{-1} \frac{\cosh (\pi y/2W)}{\cosh \{(V/V_0)\cosh^{-1}[\cosh (\pi y/2W)/\cos (\pi a_0/2W)]\}} \quad (1)$$

where the initial voltage, V_0, is measured at known crack length a_0 and the remaining geometric parameters are defined in Fig. 1. The accuracy of this approach and its applicability to both compact and single edge-notched bend specimens have been demonstrated [3] and reconfirmed by the author.

Crack-Tip Opening Displacement

The CTOD is computed from a record of load versus crack surface displacement. The crack surface displacement appropriate for analysis of a com-

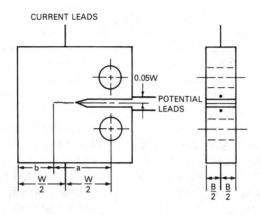

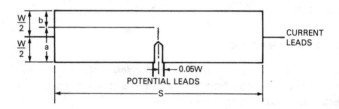

FIG. 1—*Test configuration for the* (a) *compact tension and* (b) *single-edge-notched bend specimens.*

pact specimen is the load-line displacement, while a single edge-notched bend specimen requires crack mouth displacement. Then, using the approach of BS 5762, "Methods for Crack Opening Displacement Testing"

$$\delta = \frac{K^2}{2\sigma_0 E'} + \frac{0.4b}{a + 0.4b} v_p \tag{2}$$

where

$E' = E$ for plane stress or $E' = E/(1 - \nu^2)$ for plane strain,
σ_0 = flow stress defined to be the mean of the yield and ultimate tensile stress,
E = modulus of elasticity,
K = stress-intensity factor computed from Ref *4*

$$K = \frac{P}{BW^{1/2}} \frac{(2 + \alpha)}{(1 - \alpha)^{3/2}} (0.886 + 4.64\alpha - 13.32\alpha^2 + 14.72\alpha^3 - 5.6\alpha^4)$$

$$\tag{3a}$$

for the compact specimen, or

$$K = \frac{PS}{BW^{3/2}} \frac{3\alpha^{1/2}[1.99 - \alpha(1 - \alpha)(2.15 - 3.93\alpha + 2.7\alpha^2)]}{2(1 + 2\alpha)(1 - \alpha)^{3/2}} \quad (3b)$$

for the bend specimen, where α is the relative crack depth, a/W, and v_p is the plastic component of the crack surface displacement which, with reference to Fig. 2, may be written as

$$v_p = v - P\lambda \quad (4)$$

where λ is the compliance determined from the initial linear portion of the load-displacement record.

Thus, computation of the CTOD requires knowledge of the load versus crack surface displacement record as well as the current crack length.

J-Integral

The parameter J_M [5] is computed by

$$J_M = J_D + \Delta J \quad (5)$$

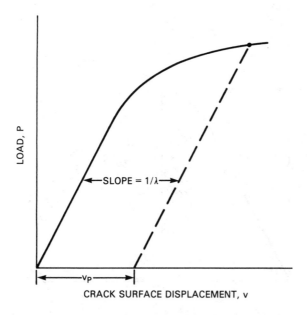

FIG. 2—*Load-displacement record analyzed to determine* δ.

where the terms J_D and ΔJ are evaluated incrementally from the load versus crack surface displacement record by

$$J_{D_{i+1}} = [J_{D_i} + (\eta/b)_i(\beta_i A_{i,i+1}/B)][1 - (\gamma/b)_i(a_{i+i} - a_i)] \qquad (6)$$

$$\Delta J_{i+1} = \Delta J_i + (\gamma/b)_i(J_{D_i} - K_i^2/E')(a_{i+i} - a_i) \qquad (7)$$

where $b = W - a$, the remaining ligament dimension, $A_{i,i+1}$ is the incremental area under the load-displacement curve as shown in Fig. 3, and, for the compact specimen

$$\beta = 1$$
$$\eta = 2 + 0.522\, b/W \qquad (8a)$$
$$\gamma = 1 + 0.76\, b/W$$

while for the bend specimen

$$\beta = W/(a + 0.4b)$$
$$\eta = 2 \qquad (8b)$$
$$\gamma = 1$$

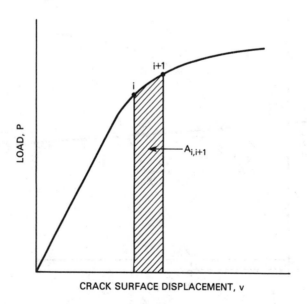

FIG. 3—*Load-displacement record analyzed to determine J.*

The term β is a geometric parameter which transforms crack surface displacement to load-point displacement by assuming rigid rotation of the crack surfaces about a point a fixed relative distance (equal to $0.4b$) ahead of the crack tip [6].

Thus, computation of J_M and the more commonly applied J_D requires knowledge of the load versus crack surface displacement record as well as the current crack length.

Additional Fracture Parameters

It is apparent, from the foregoing discussion, that the determination of δ, J_M, or J_D and the resulting R-curves requires triads of load, crack surface displacement, and electric potential data. Then, any additional parameter appropriate for fracture toughness characterization that is computed from these quantities may be also easily determined.

Plasticity-corrected stress-intensity factors are calculated by adding a plastic zone correction to the physical crack length and using this effective crack length in Eq 3 to compute K_R. An alternative approach is to use $K_J = \sqrt{JE'}$ as a plastic zone corrected stress-intensity factor.

Fracture toughness parameters such as K_{Ic} and J_{Ic} may be computed using the analyses described in the ASTM Test Method for Plane-Strain Fracture Toughness of Metallic Materials (E 399-83) or the ASTM Test Method for J_{Ic}, A Measure of Fracture Toughness (E 813-81). New parameters and analysis procedures may be easily accommodated as they are developed by adding a few program lines in the analysis software. Thus, a single fracture test can be conducted to obtain any of a number of desired fracture toughness characterizations.

Hardware Considerations

A judicious selection of hardware is essential to the establishment of a system and procedure which results in high accuracy, resolution, and repeatability. The test specimen and instrumentation cluster are shown schematically in Fig. 4. Twelve-bit analog-to-digital conversion, providing 5-mV signal resolution, is used to acquire load and displacement data. The amplified load signal, 10 V full scale, provided by the test machine yields load resolution of 0.05% full scale. The amplified displacement signal is 4 V at the full scale of 5.08 mm. Thus, the displacement resolution is 6.35 μm. These resolutions are sufficient for the accurate determination of quantities computed from the load-displacement record.

The most sensitive measurement is electric potential. For a typical fracture specimen with an initial relative flaw depth $a_0/W = 0.5$ and an applied current resulting in an initial potential $V_0 = 200$ μV, a crack extension of 25 μm

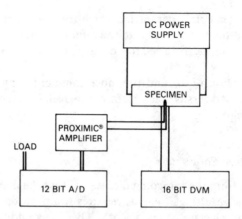

FIG. 4—*Test specimen and instrumentation cluster.*

results in an increase in the electric potential of only 0.1 μV. Thus, a 16-bit digital voltmeter is required to provide the necessary resolution.

Signal noise often becomes the factor which limits effective resolution when dealing with such low level signals; thus, care must be exercised to reduce pickup in the potential leads. However, an additional source of noise is the power supply. A typical constant current source may produce a ripple of 50 mA. This would result in an apparent peak-to-peak noise, which depends on the resistivity of the test specimen, of 0.5 μV for a typical steel specimen. Thus, selection of a power supply requires significant attention to the appropriate specifications.

The remaining required hardware is less restrictive in requirement and selection. The computer must have sufficient speed and memory to execute the required programs, support graphics programming for effective display of data and results, and be easily interfaced with a wide selection of peripherals and instrumentation.

A schematic of the computer hardware and instrumentation cluster is shown in Fig. 5. The disk drive provides rapid storage of data and results; the digital plotter complements CRT graphics, which may be dumped to the printer; the graphics tablet allows digitizing photographs of fracture surfaces for crack length measurement; and the relay actuator activates the start, hold, and reset controls of the function generator for computerized test control. Such a configuration permits a sophisticated test procedure to be developed and easily implemented.

Procedure

The procedure developed for R-curve testing and analysis may be summarized as follows:

(1) Enter test information and perform test.

(2) Determine the through-thickness-average initial and final crack lengths.

(3) Identify the point of crack growth initiation and complete analysis.

Software has been written to accomplish these tasks with minimum manual intervention. General information, specimen dimensions, and test parameters are entered with the computer keyboard in response to prompts for information. Graphics are established to provide real-time load versus crack surface displacement and R-curve displays on the CRT and potential increase versus crack surface displacement on the plotter. A test information summary and data table are established on the printer. Disk files are created for data storage and instrumentation prepared for data acquisition. The operator may then begin the test. Triads of load, crack surface displacement, and electric potential data are acquired, stored, and printed. The apparent crack extension is computed by Eq 1 where the net potentials, the difference of the measured potentials, and the zero current or thermal potential are used. The fracture parameters are computed and the graphics updated. The computer terminates the test if the load rapidly reduces signifying instability, a transducer limit is attained, or the disk data file is full. The operator may also pause or terminate the test at any time.

Subsequent to testing, the final crack length is marked by heat tint or fatigue and the specimen fractured. The fracture surface is photographed and the print mounted on the graphics tablet. Typical photographs, shown in Fig. 6, demonstrate that sufficient contrast is achieved to accurately mark the flaw borders. The initial and final flaw borders are traced and the nine-point through-thickness-average crack lengths computed for incorporation in the final analysis.

The point of initiation of crack extension is determined using a graph of electric potential increase versus crack surface displacement, as shown in Fig. 7. The start of the rapid increase in the electric potential coincides with

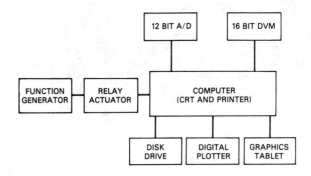

FIG. 5—*Computer hardware and instrumentation cluster.*

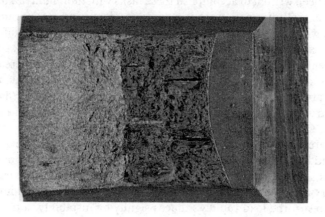

(b)

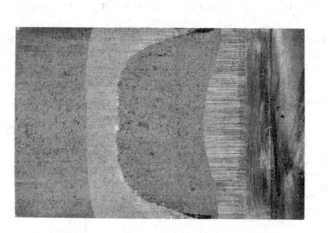

(a)

FIG. 6—Typical fracture surface for (a) 7075-T651 aluminum and (b) side-grooved HY-80 steel specimens.

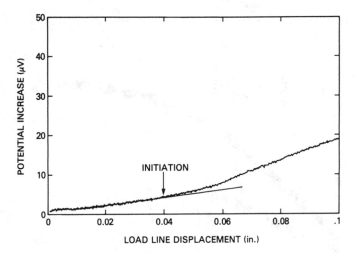

FIG. 7—*Electric potential blunting line for determining point of crack growth initiation.*

the initiation of crack growth [7]. A computer-assisted procedure displays the plot on the CRT, establishes the "blunting line," and allows the operator to identify the initiation point using a light pen. Experience has demonstrated this approach to be simple, reliable, and reproducible.

Crack lengths are computed using Eq 1 where V_0 is the potential at the point of initiation and a_0 the corresponding through-thickness-average initial crack length. A linear correction is employed to ensure that the final crack length computed from the electric potential data is equal to the final crack length measured on the fracture surface [8]. Then, the crack extension is

$$\Delta a = \Delta a_p \frac{\Delta a_f}{\Delta a_{pf}} \tag{9}$$

where the subscripts p refer to electric potential computations and f to final conditions. The ratio $\Delta a_f / \Delta a_{pf}$ has been observed to depend on the extent of crack front curvature and thus on material, specimen thickness, and side grooving conditions.

The fracture parameters are computed as previously described. Plane-stress conditions are assumed for smooth specimens while plane-strain conditions are assumed for side-grooved specimens. The data are edited such that points on the R-curves are arbitrarily spaced at least 25 μm of crack extension apart. The results are printed in both tabular and graphical form.

Results and Conclusions

Typical replicate R-curves obtained by the technique are shown in Figs. 8 and 9. Figure 8 presents δ-R curves for compact specimens of 2024-T351 and

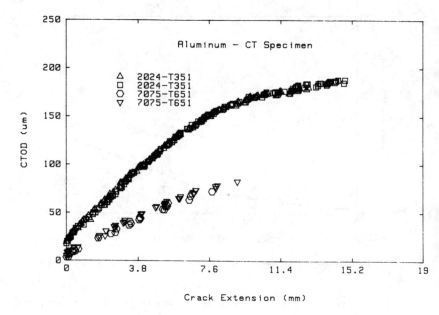

FIG. 8—δ-R *curves for aluminum alloys.*

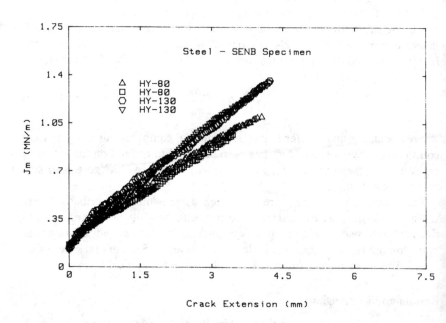

FIG. 9—J-R *curves for HY steels.*

7075-T651 aluminum alloys while Fig. 9 presents J-R curves for single edge-notched bend specimens of HY-80 and HY-130 steels. It is observed that little scatter is exhibited within each result and that the results are reproducible. Note that since the point of crack growth initiation is used as the reference point in the analysis, a blunting line is eliminated from the R-curve providing a relation between the fracture parameter and true physical crack extension. Thus, an automated technique for R-curve testing and analysis has been established. The procedure is simpler than conventional techniques by requiring only monotonic loading, minimizes required manual intervention, offers greater accuracy and resolution, permits precise determination of the initiation toughness, and allows a single fracture test to be utilized for numerous fracture toughness characterizations.

References

[1] Aronson, G. H. and Ritchie, R. O., "Optimization of the Electrical Potential Technique for Crack Growth Monitoring in Compact Test Pieces Using Finite Element Analysis," *Journal of Testing and Evaluation*, Vol. 7, No. 4, 1979, pp. 208-215.

[2] Johnson, H. H., "Calibrating the Electric Potential Method for Studying Slow Crack Growth," *Materials Research and Standards*, Vol. 5, No. 9, 1965, pp. 442-445.

[3] Schwalbe, K. H. and Hellmann, D., "Application of the Electrical Potential Method to Crack Length Measurements Using Johnson's Formula," *Journal of Testing and Evaluation*, Vol. 9, No. 3, 1981, pp. 218-221.

[4] Srawley, J. E., "Wide Range Stress Intensity Factor Expressions for ASTM E 399 Standard Fracture Toughness Specimens," *International Journal of Fracture*, Vol. 12, No. 3, 1976, pp. 475.

[5] Ernst, H. A., "Material Resistance and Instability Beyond J-Controlled Crack Growth," *Elastic-Plastic Fracture, ASTM STP 803*, American Society for Testing and Materials, Philadelphia, Vol. I, 1983, pp. 191-213.

[6] Sumpter, J. D. G. and Turner, C. E., "Method for Laboratory Determination of J_c," *Cracks and Fracture, ASTM STP 601*, American Society for Testing and Materials, Philadelphia, 1976, pp. 3-18.

[7] Lowes, J. M. and Fearnehough, G. D., "The Detection of Slow Crack Growth in Crack Opening Displacement Specimens Using an Electrical Potential Method," *Engineering Fracture Mechanics*, Vol. 3, No. 2, 1971, pp. 103-108.

[8] Saxena, A., "Electrical Potential Technique for Monitoring Subcritical Crack Growth at Elevated Temperatures," Westinghouse R&D Center Paper 79-1D3-EVPLA-P1, Pittsburgh, PA, 1979.

Timo Saario,[1] *Kim Wallin,*[1] *Heikki Saarelma,*[1]
Aki Valkonen,[2] *and Kari Törrönen*[3]

A Computer-Interactive System for Elastic-Plastic Fracture Toughness Testing

REFERENCE: Saario, T., Wallin, K., Saarelma, H., Valkonen, A., and Törrönen, K., **"A Computer-Interactive System for Elastic-Plastic Fracture Toughness Testing,"** *Automated Test Methods for Fracture and Fatigue Crack Growth, ASTM STP 877,* W. H. Cullen, R. W. Landgraf, L. R. Kaisand, and J. H. Underwood, Eds., American Society for Testing and Materials, Philadelphia, 1985, pp. 260–268.

ABSTRACT: A computer-interactive fracture toughness testing system is described. The system reliability is shown to meet the requirements of the proposed standard test procedure. Exemplary test results for compact testing, round compact testing, and three-point bend specimen geometries are presented. Based on the experiences gained, some minor additions are proposed to the standard test procedure.

KEY WORDS: fracture toughness testing, elastic-plastic fracture toughness, unloading compliance, fracture resistance curves

Design against fracture, be it brittle or slow-stable, has become or is becoming mandatory in offshore, arctic, and nuclear technology. This has resulted in increasing demand for fracture mechanics testing capacity.

As a measure of materials resistance against slow-stable crack growth, the plane-strain J_I-R curve [1] is probably the most widely accepted one. In measuring a J_I-R curve, the main experimental problem is to reliably detect the amount of crack extension during the test. The single-specimen elastic compliance method described in the tentative test procedure for plane-strain J_I-R curve (Albrecht et al [1]), was found in a round-robin test program [2] to be satisfactory in this respect.

[1]Research officers, Metals Laboratory, Technical Research Center of Finland, Espoo, Finland.
[2]Research officer, Academy of Finland, Espoo, Finland.
[3]Head, Materials Technology Section, Metals Laboratory, Technical Research Center of Finland, Espoo, Finland.

The elastic compliance method, besides being relatively easily automated, has the additional advantage of making possible the study of the influence of macroscopic material inhomogeneities on fracture resistance. This is because a complete J_I-R curve can be extracted from one specimen, and thus differences in J_I-R curves reflect variations in macroscopic material properties.

In this paper a computer-interactive testing system based on the single-specimen elastic compliance method is described. Some suggestions are made regarding the use of the recommended [1] pin-loaded compact and three-point bend specimens as well as the round compact and precracked Charpy specimens.

Description of the Testing System

The block diagram of the testing system is shown in Fig. 1. The measurement-control loop is as follows. The computer triggers the function generator via a relay actuator to send an increasing, decreasing, or constant signal to the control circuit of the servohydraulic testing machine. As the test is performed under clip gage feedback control, the servo valve is then adjusted so as to have equal output signals from the clip gage and the control circuit. The resulting changes in the load and displacement signals are recorded by a digital voltmeter. This measurement is again controlled by the microcomputer through a multichannel controller unit. All data are stored on a magnetic floppy disk.

During the test, plotters are used to show graphically the load-displacement curve and the J_I-R curve as the test proceeds. These plots are used as a basis for decision-making during the test. The decisions are then carried out using preprogrammed special function keys of the computer. The decisions can also be carried out automatically by the computer.

The testing system as such conforms in detail to the requirements of the ASTM Test Procedure for J_{Ic}, a Measure of Fracture Toughness (E 813-81) as well as to those recommended by Albrecht et al [1].

The specimen size can vary freely between 10- and 100-mm thickness and the test can be performed at any temperature from $-100°C$ to $+300°C$.

Experiences on Various Specimen Geometries

Compact Testing Specimens

Figure 2 shows a comparison of the physical crack extension and that predicted by the above-described testing system for some 30 tests. The tests were performed with 25-mm-thick compact testing (CT) specimens at various temperatures. As seen, the results lie rather well between the broken lines, which represent the $\pm15\%$ relative maximum error allowed [1] for a test to be quali-

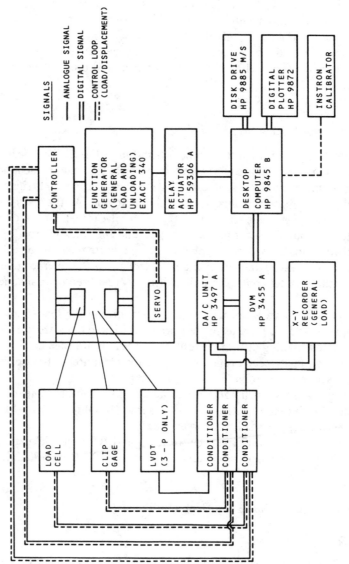

FIG. 1—*Block diagram of the testing system.*

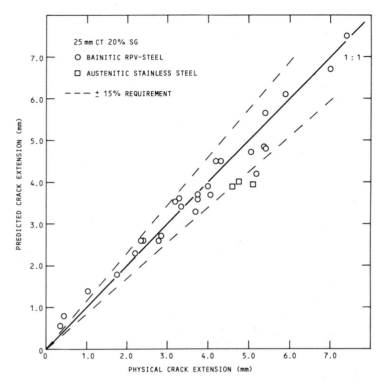

FIG. 2—*Comparison of predicted and measured values of final crack extension. Broken lines show the maximum error allowed by the standard test procedure.*

fied as valid. Thus, at least for CT specimen geometry, the recommended test practice leads to acceptable results.

The recommended test practice [1] gives very broad limits for the loading rate to be used for initial crack length measurements. Also, during the actual test, unloading rates as slow as desired can be used. In our experience, changing the unloading rate can in some cases result in marked differences in the measured compliance. For a typical 25-mm-thick CT specimen, the recommended minimum and maximum loading rates for initial crack length measurements are about 0.02 and 2 mm/min, respectively. Figure 3 shows the results of repeated initial crack length measurements when different loading rates were used with a nearly constant unloading load drop range. Figure 4 shows the corresponding effect of the loading rate on the linear regression analysis correlation coefficient. During every unloading a total of 40 load-displacement points were used for the compliance calculation, and five unloadings were performed with each loading rate. As seen, the scatter is largest for the lowest loading rate, although the requirement of the tentative test procedure is met for all loading rates. It seems that for initial crack length deter-

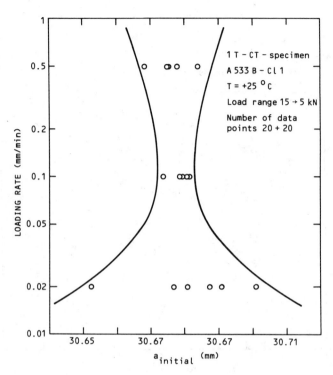

FIG. 3—*Effect of loading rate on the predicted initial crack length of a 25-mm CT specimen.*

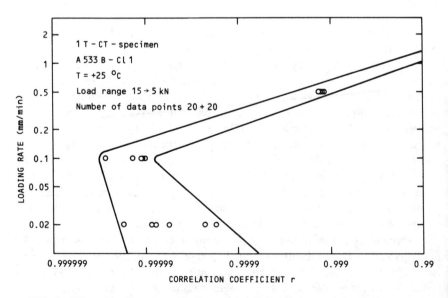

FIG. 4—*Effect of loading rate on the linear regression analysis correlation coefficient* r.

minations the loading rate does not have a major effect on the calculated compliance. However, it seems justifiable to suggest that a requirement of constant loading rate during the test should be added to the tentative test procedure.

Round Compact Testing Specimen Geometry

The round compact testing (RCT) specimen geometry shown in Fig. 5 is slightly different from the one discussed by Newman [3]. The present geometry with machined sharp edges at specimen load line is better suited for J_I-R curve testing with irradiated specimens. Finite-element calculations showed [4] that the compliance of the present geometry was only slightly different from that given by Newman. The resulting compliance curve for $0.4 \leq a/W \leq 0.85$ is of the form

$$a/W = 1.015 + 3.206 \cdot U_x - 43.986 \cdot U_x^2 + 83.179 \cdot U_x^3$$

where $U_x = 1/[\text{LOG(BEV}/P) + 1]$, and V/P is the elastic compliance.

Besides establishing a sound crack length-compliance calibration curve, the only other difficulty encountered in testing RCT specimens has been the limited space reserved for the clip gage (see Fig. 5). In our experience, a clip gage touching the notch face even slightly results in very abrupt changes in the measured compliance during the test. As these changes are nicely repeatable, it should not take too long to figure out where the problem is.

As shown in Fig. 6, the J_I-R curves measured with a 12.5-mm RCT specimen coincide well with the ones measured with a 25-mm CT specimen.

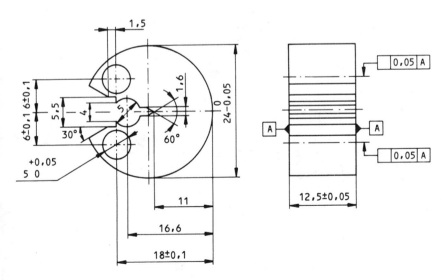

FIG. 5—*Round compact testing (RCT) specimen dimensions.*

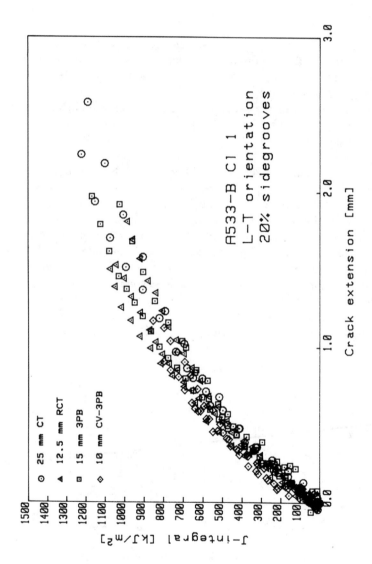

FIG. 6—Comparison of J_r-R curves measured with four different specimen geometries at room temperature. The results of two tests are shown for each geometry.

Three-Point Bend Specimen Geometry

Three-point bend (3PB) specimen geometry had advantages over CT and RCT geometries in that interpretation of the test record in terms of the J-integral is on a much more sound basis. However, from the experimental point of view, the 3PB geometry has some serious disadvantages. Possible plastic indentation at load points gives uncertainty to the measured load-point displacement and accordingly to the calculated J-value. Freely moving roller pins, on the other hand, allow for possible changes in the span width during the test, which then induces errors in the crack length prediction.

Tests were conducted with both the recommended 3PB geometry [1], using a thickness of 15 mm, and the Charpy-V specimen geometry. During both tests, the span width was measured after every unloading using a device the accuracy of which was ±0.10 mm. In addition to the load-point displacement, the crack-opening displacement at the front face of the specimen was measured with a clip gage. The crack-opening displacement was used to calculate the crack length from the compliance curve given by Joyce et al [5]. Possible errors in J-values produced by plastic indentation were not corrected for.

The resulting J_1-R curves are shown in Fig. 6. No significant geometry dependence of J_1-R curves was found. During the 15-mm 3PB specimen tests, the 12.5-mm-diameter roller pins drifted outward almost 1.5 mm each, but as the contact points moved inward exactly the same amount, the span width remained constant to a high degree.

The J_1-R curve derived from the Charpy-V specimen differs somewhat from all the others. In this case, using 8-mm-diameter roller pins, the span width first remained constant to a load-point displacement of about 2.0 mm, and then steadily decreased from 39.7 to 38.3 mm.

It should be noted that no "apparent negative crack growth" has been found when the crack length has been calculated based on the crack-opening displacement. Neale [6], using a compliance calibration curve based on load-point displacement, found negative crack growth extending up to about 0.5 mm for a Charpy-V specimen. The use of a clip gage for compliance measurement and a separate displacement gage for measuring the load-line displacement, as also recommended by Albrecht et al [1], offers no indication of negative crack growth.

Summary and Conclusions

Automated testing systems provide an adequate solution to the problem of increasing demand for advanced materials testing. For the particular case of J_1-R curve testing, the single-specimen elastic compliance method has been most thoroughly investigated and found reliable. In this paper, one successfully operating testing system utilizing this method has been described.

With CT and RCT specimen geometries, the recommended testing practice [1] was found detailed enough. However, a requirement of constant crosshead speed during the test should be added. With 3PB specimen geometry, such details as roller pin and loading head diameters probably control the extent of span width changes during the test. More specific limits should be given to these diameters so as to make tests performed with different test fixtures more readily comparable.

Acknowledgment

This work is a part of the Reliability of Nuclear Materials Program performed at the Technical Research Center of Finland (VTT) and financed by the Ministry of Trade and Industry in Finland. Additional financing by the Emil Aaltonen foundation (T.S.) is gratefully acknowledged.

References

[1] Albrecht, P., Andrews, W. R., Gudas, J. P., Joyce, J. A., Loss, F. J., McCabe, D. E., Schmidt, D. W. and Van Der Sluys, W. A., *Journal of Testing and Evaluation*, Vol. 10, No. 6, Nov. 1982, pp. 245–251.
[2] Gudas, J. P. and Davis, D. A., *Journal of Testing and Evaluation*, Vol. 10, No. 6, Nov. 1982, pp. 252–262.
[3] Newman, J. C., *International Journal of Fracture*, Vol. 17, No. 6, Dec. 1981, pp. 567–578.
[4] Valkonen, A., Talja, H., and Ikonen, K., "Elastic-Plastic Analysis of Fracture Mechanics Test Specimens. Part 1," Technical Research Center of Finland, Research Report 150, 1983 (in Finnish).
[5] Joyce, J. A., Hasson, D. F., and Crowe, C. R., *Journal of Testing and Evaluation*, Vol. 8, No. 6, Nov. 1980, pp. 293–300.
[6] Neale, B. K., *International Journal of Pressure Vessels and Piping*, Vol. 12, 1983, pp. 207–227.

David A. Jablonski[1]

Computerized Single-Specimen J-R Curve Determination for Compact Tension and Three-Point Bend Specimens

REFERENCE: Jablonski, D. A., "**Computerized Single-Specimen J-R Curve Determination for Compact Tension and Three-Point Bend Specimens,**" *Automated Test Methods for Fracture and Fatigue Crack Growth, ASTM STP 877*, W. H. Cullen, R. W. Landgraf, L. R. Kaisand, and J. H. Underwood, Eds., American Society for Testing and Materials, Philadelphia, 1985, pp. 269–297.

ABSTRACT: A computer-controlled test system was developed to measure J-R curves and J_{Ic} on compact tension and three-point bend specimens. The crack length was measured by unloading compliance. Improved crack length accuracy was obtained by amplifying the load and crack opening displacement (COD) signals by factors of 10 to 20 before the unloading. Crack length accuracy was further improved by collecting 500 data pairs on unloading using a fast data acquisition method. In the compact tension specimen, crack length and load-line displacement were measured with a load-line COD gage. In the three-point bend specimen, load-line deflection was measured using the crosshead deflection corrected for specimen and test fixture stiffness. An error analysis was performed to determine the crack length errors. The analysis showed the beneficial effect of using load and COD ranging to reduce the crack length errors.

J-R curves were measured on 17-4-Ph stainless steel heat treated to 948 MPa yield. The J-R curves were measured on two specimen geometries, compact tension and three-point bend, using four starting a/w ratios from 0.65 to 0.80 and three side groove depths, 0%, 12.5%, and 25%. The test results showed that in specimens with side grooves of 12.5% and 25%, the measured J_{Ic}-values were independent upon both specimen geometry and starting a/w ratio. The material tearing modulus, $T_{material}$, was found not to be a material property since its value depended upon specimen geometry.

KEY WORDS: J-integral, J_{Ic}, $T_{material}$, J-Resistance curve, fracture mechanics, elastic-plastic, unloading compliance, crack length error, automated test methods, computer control, 17-4-Ph stainless steel

[1]Materials scientist, Instron Corp., Canton, MA 02021.

Nomenclature

A_i	Area under load: load-line deflection curve
$A_{i,i+1}$	Area under load: load-line deflection curve between ith and $i + 1$ loadings
a	Crack length
a_0	Initial crack length
Δa_{error}	Error in crack length
Δa_p	Physical crack advance
B	Specimen thickness
B_{eff}	Effective thickness $[B - (B - B_{\text{net}})^2/B]$
B_{net}	Net thickness
b	Uncracked ligament $(W - a)$
C_m	Test machine and fixture compliance
CHD	Crosshead displacement
ds	Element of displacement along path S
E	Modulus of elasticity
E'	Plane-strain modulus $[E/(1 - \nu^2)]$
ΔE_{er}	Error in modulus measurement
$f(a_0/w)$	Dimensionless coefficient used to calculate J
J	J-integral
J_{C}	J-integral corrected for crack advance
J_{Ic}	Fracture toughness
J_{Q}	Provisional value of J_{Ic}
K	Stress-intensity factor
P	Load
ΔP_{er}	Error in load measurement
S	Span width of three-point bend fixture
S_b	Standard error estimate of slope
$S_{x/y}$	Covariance of x and y
S_x	Variance of x
S_y	Variance of y
T	Traction vector along path S
T_{material}	Material tearing modulus
u	Displacement vector
U	Transfer function for compliance
V_0	Crack opening displacement (COD)
$\Delta V_{0\text{er}}$	Error in COD measurement
W	Specimen width
w_ρ	Energy density
β	Intercept in regression analysis $(y = mx + \beta)$
δ	Load-line deflection
ν_i	Dimensionless coefficient used to calculate J
σ_y	Yield strength

σ_f Flow strength
ν Poisson's ratio
Γ Path used to calculate J-integral

The use of the J-integral in materials evaluation has become increasingly important since the original work by Eshelby [1], Hutchinson [2], and Rice [3,4] showed that the J-integral describes the stress field in a cracked body under elastic-plastic conditions. Due to the many shortfalls in the K_{Ic} test in measuring a materials resistance to static cracking, the J-integral approach was applied to various test specimens to determine a J-integral fracture criterion. As a result of the work by many researchers a standard J_{Ic} fracture test was developed by ASTM [5]. The experimental and theoretical work that was the basis of the ASTM standard came mainly from the work in Refs 6-9. The objective of the ASTM method is to determine the value of J at the initiation of crack growth. The resistance of a material to crack growth is described by a J-R curve, which measures J as a function of physical crack advance. The J-R curves are assumed to be bilinear for small amounts of crack extension with the initial linear section defining the crack-tip blunting ($J = 2\sigma_f \Delta a$) and the second linear section defining the growth of a crack from the blunted notch. The intersection of these curves defines J_{Ic}, the fracture toughness value.

Two test techniques are used to determine the J-R curve: the multiple-specimen technique and the single-specimen technique. The single-specimen technique is potentially superior to the multiple-specimen technique since an entire J-R curve is generated with the same specimen. The single-specimen technique eliminates the effect of material variability on the J-R curve. The technique is far more efficient in the use of test material, which is generally not very plentiful. The greatest problem in the single-specimen technique is the accurate measurement of crack length. Many of these problems of obtaining accurate crack length measurements have been discussed by Clarke [8-10], Jones [11], and Hewitt [12]. The most common method to determine crack length is by unloading compliance, in which the unloading is limited to 10% of the current load to prevent crack growth by fatigue processes and any subsequent effects of the unloading on the R-curve development. In typical test machines, full-scale equals 10 V; thus the load signal available for the unloading compliance measurements is 1 V or less and the COD signal is usually smaller than the load signal. Care must be exercised to accurately measure the slope of these low-level signals or the crack length measurement will not be accurate. A practical solution to this problem is to use a computer system for data collection or test control or both. A computer-assisted single-specimen J_{Ic} test was developed by Joyce [13] using a Tektronix-based computer system in which the computer was used for data storage and analysis. A more sophisticated test system was recently developed by Van Der Sluys et al [14]. Their system contained both computer control with data storage and analysis using a Digital Electronics Corp.-based system. In this system, crack length accuracy was improved by computer-controlled ranging and zero suppression. In both computer-assisted and computer-controlled test systems, excellent results and good agreement were obtained compared with tests performed using the multiple-specimen technique.

An additional method which should improve the accuracy of crack length by the unloading compliance technique is to collect a large quantity of data points (500 to 1000) during each unloading and allow the statistics of the signal noise to improve the slope resolution. The primary objective of this study was to develop a computer-controlled single-specimen J_{Ic} test system utilizing a fast (10 ms) data acquisition method. A crack length error analysis method was developed to determine the usefulness of load and COD ranging in reducing the crack length error. The secondary objective was to measure J-resistance curves for 17-4-Ph stainless steels studying the effect of specimen geometry (three-point bend and compact tension), side groove depth, and starting crack length to width ratio. Previous work has shown varying results with respect to the effect of these parameters on J-resistance curves [7-9,15-21]. Test results were focused not only on J_{Ic}, but also on the value of material tearing modulus $T_{material}$. Published work on the tearing modulus [9,22-25] did not seem to adequately show geometry and size independence of $T_{material}$. The test results were focused on the validity of J_{Ic} as a fracture parameter and on the validity of the tearing modulus as a true material property.

Experimental Procedure

Material and Specimen Geometries

The material used in this investigation was a precipitation-hardened stainless steel 17-4-Ph. A typical chemical composition of this alloy is listed in Table 1. The alloy was obtained in the solutionized condition and was subsequently aged at 620°C for 4 h. The measured room temperature mechanical properties are listed in Table 2. Two specimen geometries were used for this investigation, the three-point bend geometry and the compact geometry. The specimens were machined according to the ASTM Test Method for J_{Ic}, a Measure of Fracture Toughness (E 813-81), with the longitudinal-transverse (L-T) orientation as described in the ASTM Test Method for Plane-Strain Fracture Toughness of Metallic Materials (E 399-83). The specimen dimensions used in these experiments are listed in Table 3. Some of the specimens

TABLE 1—*Typical composition of 17-14-Ph stainless steel (weight %).*

C	Mn	Si	Cr	Ni	Cb	Cu	Fe
0.04	0.40	0.50	16.5	4.25	0.25	3.60	bal

TABLE 2—*Room temperature mechanical properties of 17-4-Ph stainless steel.*

Yield strength	948 MPa
Ultimate tensile strength	1020 MPa
Flow stress	984 MPa
Fracture strain	14.7%
Modulus of elasticity	193 GPa
Poisson's ratio	0.27
Plane strain modulus	207 GPa

TABLE 3—*Specimen dimension for J-integral specimens.*

Specimen Type	Thickness, B, mm	Width, W, mm	Machined Notch Length, a, mm
Compact	25.4	50.8	12.7
Three-point bend[a]	12.7	25.4	6.35

[a] Span width = 101.6 mm (4w).

were side grooved with an included angle of 90 deg. Side grooves which resulted in either 12.5% or 25% reduction in thickness were used.

The COD gage was attached to the specimens by the use of knife edges machined in the specimen. Knife edges were made by a 9.5-mm hole which was drilled 4.0 mm off center. This hole then provided a 5.0-mm opening for attachment of the COD gage. In the compact specimen the COD gage was located at the load line whereas in the three-point bend specimen the COD gage was located on the surface of the specimen adjacent to the machined notch.

Computer Hardware and Test Equipment

Fatigue precracking of specimens was done on an Instron computer-controlled servohydraulic test system using the fatigue crack propagation program. Details of this equipment and procedure are described elsewhere [26]. Fatigue precracking was done using a constant-K test with a delta-K of 23 MPA$\sqrt{m}$. Fatigue precracks were extended to a/w's from 0.65 to 0.80 from an initial machined notch a/w of 0.50. The fatigue precrack procedure was performed to conform to ASTM Method E 813, Section 7.6.

The J-integral tests were done on a computer-controlled Instron Model 1125 electromechanical test system. The electromechanical test system was preferred over the servohydraulic test system because of the superior displacement control of the electromechanical test system. A block diagram of the test system is shown in Fig. 1. The zero suppression of load and COD was accom-

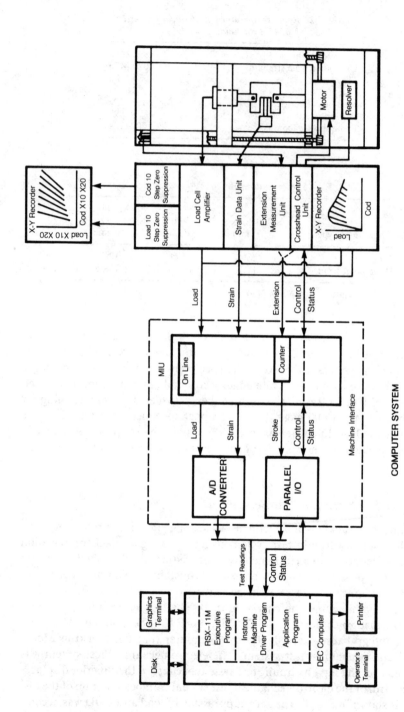

COMPUTER SYSTEM

FIG. 1—*Block diagram of computer-controlled test instrument.*

plished using a ten-step zero-suppression module. The computer system used was a DEC PDP 11/23 with 128K word memory and a floating point processor. A large disk storage capacity of 20M byte was used. The disk storage system consisted of one RL02 and two RL01 hard disks. In each J-integral test the raw test data, some 40 000 data triplets of load, COD, and load-line deflection, were stored on the disk to allow data reduction by various techniques. The load and COD signals were input to the computer system by use of a 12-bit analog-to-digital (A/D) converter. The crosshead extension was measured by use of an optical encoder whose output was measured with a digital counting circuit in the Machine Interface Unit (MIU). The extension was measured by the computer by reading a digital 16-bit word through an 80-bit parallel input/output (I/O). The parallel I/O was used for machine control and status and for the measurement of extension. The MIU was fitted with a computer-controlled variable-speed option which allowed the computer to command any crosshead velocity within the capability of the test instrument.

The crack opening displacement was measured with Instron Model 2670 COD gages. The grips used for compact specimens were Instron Model 2750 fracture mechanic grips. The grips used for three-point bend specimens were Instron Model 2810 bend fixture.

Software Description

A program called JRTEST was developed to perform real-time control, data acquisition, and data analysis for the J-R test. The test system software was designed to test either a compact specimen or a three-point bend specimen according to the single-specimen J_{Ic} test procedure using periodic unloading to determine crack length. The operating system software was RSX11M, which was designed for multitasking and multiprogramming environments. The J-R test program was written in BASIC-plus 2 (compiled BASIC) to take advantage of the flexibility of a high-level language. The Instron machine driver (IMD), which is a collection of macro-subroutines, was used to perform all of the machine control and data acquisition tasks. The IMD was built to link to the executive I/O page directly and thus receives high priority for machine control tasks.

There are four tasks included in the J-R test software: CALEM, JRTEST, J_{Ic}, and EZGRAF. The interaction of the various tasks and their use of disk files are shown in Fig. 2. The CALEM task measures computer calibration constants and stores them in a disk file. The computer calibration constants are used by the IMD to convert A/D readings to physical units of force and displacement. JRTEST is the main program which inputs the necessary test parameters, runs the real-time test, analyzes the results, and outputs the results to the terminal and disk files. The J_{Ic} task takes the stored reduced data of (J, Δa_p) and calculates J_{Ic}, $T_{material}$, and checks the validity of the test

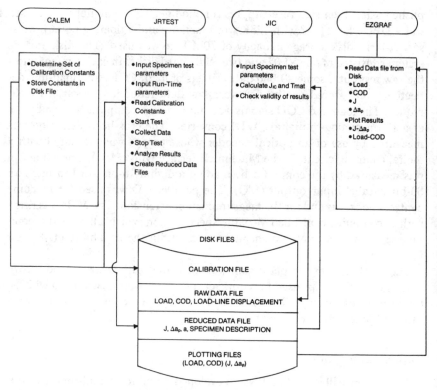

FIG. 2—*Interaction of program tasks and data files for automated* J_{Ic} *test system.*

results. The EZGRAF task plots the test results of both $(J, \Delta a_p)$ and (Load, COD) on Tektronix plotters.

A simplified flow diagram of JRTEST is shown in Fig. 3. The brackets on the side of the diagram indicate the function of each segment. The program can be divided into three sections; test parameter input, real time control with data acquisition, and data analysis. During the real-time control section, there are many simultaneous checks performed to assure the test is running properly. The software constantly monitors the feedback signals and will stop the machine if they are greater than full-scale values. During the monotonic loading section the COD is constantly monitored and the test is stopped when the COD passes the prescribed COD value. During the monotonic unloading the load is constantly monitored and the test machine stopped when the load reaches the lower load value. In either case in the monotonic loading or unloading sections, data are constantly collected at either 10- or 20-ms intervals and stored in a disk file. The data collected consisted of load, COD, and load-line deflection. The load-line deflection for the three-point bend specimen was obtained by the following equation

$$\delta = \text{CHD} - C_m \cdot P \tag{1}$$

The load-line deflection for the compact specimen was equal to the COD since the COD was measured at the load line. The value of C_m was measured using an uncracked specimen. The C_m-value was a function of the load cell capacity used for the test. The higher the load cell capacity, the lower the C_m value. For this experiment the value of C_m was 1.496×10^{-5} mm/N for the 22 240 N capacity load cell and 1.13×10^{-5} mm/N for the 88 960 N capacity load cell. A drop in the load was observed when the specimens were held at constant displacement near maximum load. This load relaxation has been reported to cause a substantial error in the calculated crack length if incorporated into the unloading [9,14]. The unloading was delayed until the load relaxation rate was less than a user-specified value (default 2.0 N/min). This delay was accomplished by software data collection and analysis loop.

In many of the specimens it was desirable to measure the J-R curve for large crack advances ($\Delta a_p > 2.5$ mm). This caused some problem in that generally by the time the crack had propagated 2.5 mm, the linear range of the COD gage had been used up. This problem was solved by changing the COD gage to one with a larger gage length and having the software keep track of the absolute value of the COD and load-line displacement.

It was felt that a real-time data analysis was unnecessary, so the raw data were stored on a disk file, making it possible to analyze the same data many times using different techniques.

Analysis of Raw Test Data

The details of the J-integral can be found in Refs 2–4 and 27. The J-integral is a path-independent line integral which may be calculated by using any path which encompasses the crack tip. The expression for J-integral is

$$J = \int_\Gamma W_\rho dy - \bar{T} \frac{\partial u}{\partial x} ds \tag{2}$$

The expressions used to approximate J for the compact tension and three-point bend specimens for $a/w > 0.50$ are

$$J = A/Bb \cdot f(a_0/w) \tag{3}$$

for three-point bend

$$f(a_0/w) = 2.0$$

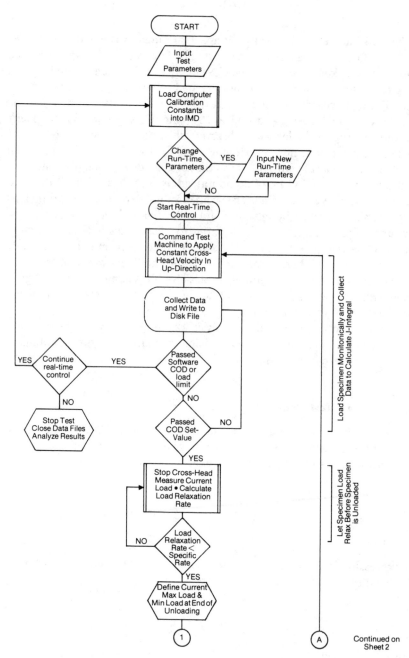

FIG. 3—*Flow diagram of automated J-R test program.*

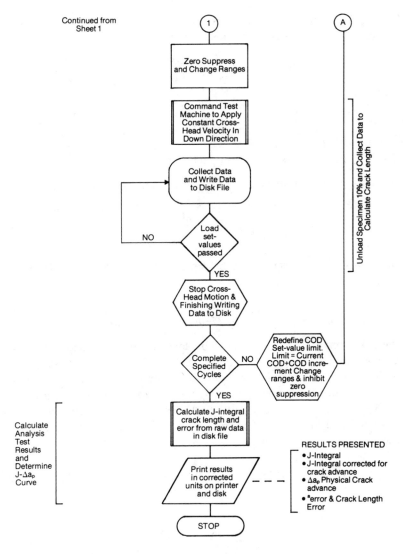

FIG. 3—*(continued)*.

for compact tension specimen [4,27]

$$f(a_0/w) = 2[(1 + \alpha)/(1 + \alpha^2)]$$
$$\alpha = [(2a_0/b)^2 + 2(2a_0/b) + 2]^{1/2} \qquad (4)$$
$$2a_0/b = 2a_0/(w - a_0)$$

The expression for J, Eq 3, does not account for physical crack advance. J is corrected for physical crack advance by [4,10]

$$J_{i+1} = \left[J_i + \left(\frac{f(a/w)}{b} \right)_i \frac{A_{i,i+1}}{B} \right] \cdot \left\{ 1 - \left(\frac{\gamma}{b} \right)_i [(a_p)_{i+1} - (a_p)_i] \right\} \quad (5)$$

where

$\gamma_i = 1$ for three-point bend, and
$\gamma_i = 1 + 0.76(w - a/w)$ for compact tension.

The tearing modulus of the material was defined by the following equation

$$T_{material} = \frac{\partial J}{\partial a} \frac{E}{\sigma_f 2} \quad (6)$$

The slope of the J-integral curve between $(J/2\sigma_f + 0.15)$ and $(J/2\sigma_f + 1.5)$ was used to define $\partial J/\partial a$.

Determination of Crack Length by Unloading Compliance

The crack length in a specimen can be determined by elastically unloading a specimen and measuring the compliance V_0/P. Saxena and Hudak [28] have determined polynomial relationships between compliance and crack length for compact tension, wedge opening loading (WOL), and center-cracked specimens. Jablonski [29] has determined similar relationships for three-point bend specimens. The crack length is determined by the following relationship

$$a/W = C_0 + C_1 U + C_2 U^2 + C_3 U^3 + C_4 U^4 + C_5 U^5 \quad (7)$$

where

$$U = 1 \bigg/ \left[\left(\frac{E' V_0 B_{eff}}{P} \right)^{1/2} + 1 \right] \quad (8)$$

$$B_{eff} = \left[B - \frac{(B - B_{net})^2}{B} \right] \quad (9)$$

The coefficients used in Eq 7 are listed in Table 4. The effective thickness, B_{eff}, was used in place of the specimen thickness, B, based upon the results obtained of Shih and deLorenzi [30] on side-grooved specimens. The compliance V_0/P was found by a linear regression analysis of the unloading load-COD data. In this analysis, only part of the unloading data was used to deter-

TABLE 4—*Crack length coefficients for various geometries.*

Geometry	C_0	C_1	C_2	C_3	C_4	C_5
Compact[a]	1.0002	−4.06319	11.242	−106.043	464.335	−650.677
Three-point bend[b]	0.994516	−3.6925	−1.70627	36.472	−106.443	125.51

[a] COD measured at load line
[b] COD measured at surface.

mine the slope because of nonlinear behavior due to transducer hysteresis. Compliance calculations were done using the slope of the unloading curve between the upper and lower limits. The limits were defined as percentages of the maximum and minimum load. Typical limits were 70% and 15%. During each unloading 500 to 1000 data pairs were collected in order to determine the slope V_0/P accurately.

Calculation of J_{Ic} and Validity Checks

The value of J_{Ic} was determined according to the methods described in ASTM Method E 813. The R-curve was obtained by a linear regression curve of qualified data. The following validity checks were performed

$$B > 25 \, J_Q / \sigma_f$$

$$B_{net} > 25 \, J_Q / \sigma_f$$

$$(W - a_0) > 25 \, J_Q / \sigma_f$$

$$\frac{dJ}{da} < \sigma_f$$

The data used to calculate the R-curve were checked such that B, b, and B_{net} were greater than $15J/\sigma_f$; data with values lower than $15J/\sigma_f$ were rejected for the R-curve analysis.

Results

A series of experiments was conducted to evaluate the influence of a variety of parameters on the J-resistance curves of 17-4-Ph stainless steel. The parameters studied included starting a/W ratio, side groove depth, and specimen geometry. A typical plot of load versus COD for these experiments is shown in Fig. 4. The COD increment was fixed at the beginning of the experiment and unloading was always 10% of the current load. The COD increment was chosen so that a minimum of eight qualified data points would be obtained for determining the R-curve.

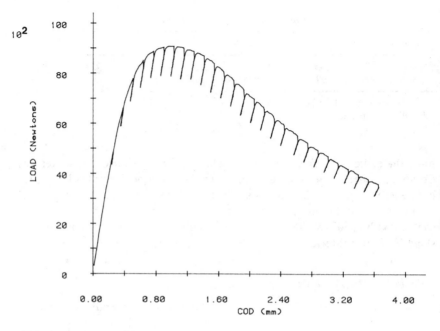

FIG. 4—*Load versus COD for three-point bend specimen with* a/W = 0.70, 12.5% *side grooves.*

The effect of side groove depth on the J-resistance curves is shown in Figs. 5–7. In the J-resistance plots the dashed line represents the calculated blunting line from the expression $J = 2\sigma_f\Delta a$. The side groove depth had very little effect on the J-resistance curves for compact tension specimens. The side groove depth had a substantial influence on the J-resistance curves in the three-point bend specimens. In the bend specimens the J-resistance curves with no side grooves had a substantially steeper slope than those for specimens with side grooves of 12.5% and 25%. In the bend specimens the J-resistance curves were identical for specimens with 12.5% and 25% side grooves.

The effect of starting a/W ratio on the J-resistance curves is shown in Figs. 8–11. In Figs. 8 and 9 the data for specimens with 25% side grooves are plotted and in Figs. 10 and 11 the data for specimens with 12.5% side grooves are plotted. There are no consistent trends in the data to correlate with the starting a/W ratio. In Fig. 10 there appears to be an ordering of the curves with respect to decreasing J with increasing a/W. An examination of the data showed the test with $a/W = 0.80$ was not valid, failing the criteria that $(W - a) > 25J/\sigma_y$.

In Fig. 11 the data for $a/W = 0.65$ were considerably different than those obtained at a/W ratios of 0.70, 0.75, and 0.80. A reason for the change in the J-resistance curve at $a/W = 0.65$ has not been found, but it is suspected that this difference was caused by a material variation.

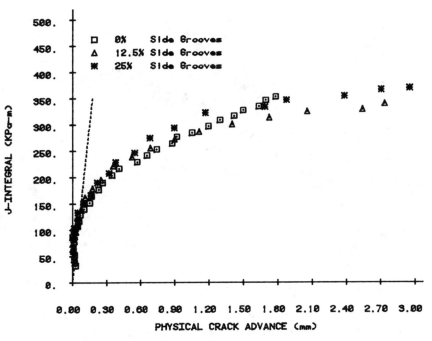

FIG. 5—*Effect of side grooves on the J-R curve with CT geometry, a/W = 0.70.*

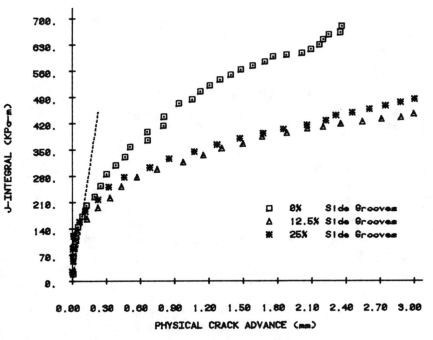

FIG. 6—*Effect of side grooves on J-R curve with bend geometry, a/W = 0.70.*

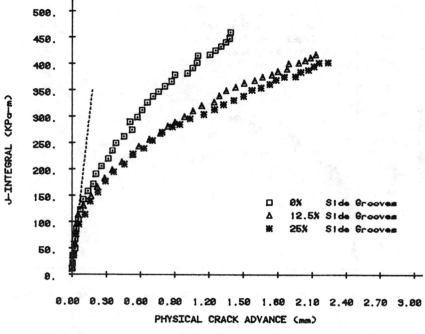

FIG. 7—*Effect of side grooves on J-R curve with bend geometry*, a/W = 0.80.

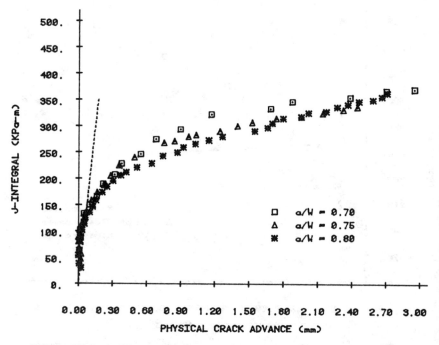

FIG. 8—*Effect of* a/W *ratio on the J-R curve of the CT geometry, 25% side grooves.*

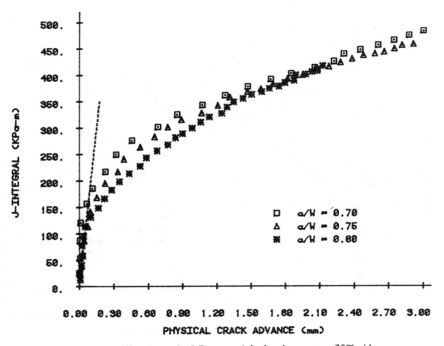

FIG. 9—*Effect of* a/W *ratio on the J-R curve of the bend geometry, 25% side grooves.*

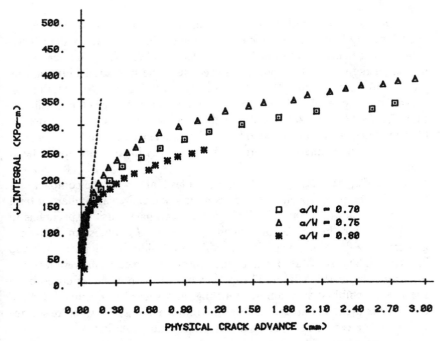

FIG. 10—*Effect of* a/W *ratio on the J-R curve of the CT geometry, 12.5% side grooves.*

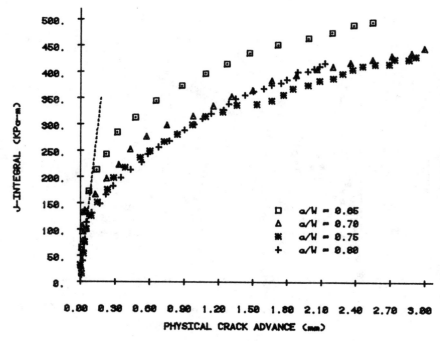

FIG. 11—*Effect of* a/W *ratio on the J-R curve of bend geometry, 12.5% side grooves.*

Comparisons were made between the measured blunting line and the calculated blunting line. These comparisons showed that in all cases the experimental and calculated blunting line coincided with the three-point bend specimens. In the compact tension (CT) specimens the measured blunting line always had a greater slope than the calculated blunting line. An examination of J-resistance curves of CT specimens that have been published in the literature [7,8,11,13,15,16,21] has shown that they all have an initially steeper blunting line than that calculated by $J = 2\sigma_f \Delta a$. The calculated blunting line appears only to estimate the blunting line correctly for the three-point bend specimens.

Values of J_{Ic} and $T_{material}$ are listed in Tables 5 and 6 for the compact tension and three-point specimens. The average value of J_{Ic} was similar for both geometries, varying from 195 kPa-m for the compact tension specimens to 205 kPa-m for the three-point bend specimens. J_{Ic}-values for non-side-grooved three-point bend specimens exhibited a wide variation, from 205 to 283 kPa-m. The J_{Ic}-values for non-side-grooved CT specimens were less varied, from 180 to 193 kPa-m. The crack front in the non-side-grooved bend specimens exhibited substantial tunneling, and shear lips occupied 35% of the fracture area. The stress state was not uniform across the fracture surface with only the center 65% being under plane-strain conditions. $T_{material}$ was

TABLE 5—*J-R data on compact tension 17-4-Ph stainless steel.*

Specimen No.	% Side Grooves	a_i/W	J_{Ic}, kPa $\cdot$ m	$T_{material}$
5517-4-1	0	0.65	193	27.34
5517-4-2	0	0.70	180	23.00
5517-4-3	12.5	0.70	195	18.56
5517-4-6	12.5	0.75	231	17.18
5517-4-4	12.5	0.80	170	18.54
5517-4-7	25	0.70	208	20.00
5517-4-9	25	0.75	209	16.31
5517-4-8	25	0.80	190	16.20

Avg J_{Ic} = 194.5 ±18.5.
Avg $T_{material}$ = 17.8 ±1.5 (see Note 1).
Note 1: does not include data for 0% side-grooved specimens.

TABLE 6—*J-R data on three-point bend 17-4-Ph stainless steel.*

Specimen No.	% Side Grooves	a_i/W	J_{Ic}, kPa $\cdot$ m	$T_{material}$
5517-4-16	0	0.70	283	50.28
5517-4-17	0	0.80	204.7	44.26
5517-4-24	12.5	0.65	264	28.10
5517-4-22	12.5	0.70	215	24.30
5517-4-23	12.5	0.75	195	23.65
5517-4-25	25	0.80	160.4[a]	30.96
5517-4-20	25	0.70	237	22.52
5517-4-21	25	0.75	193	27.80
5517-4-19	25	0.80	174	26.53

[a] Not valid W-a_i criteria not met.
Avg J_{Ic} = 205.5 ±36.0.
Avg $T_{material}$ = 26.27 ±2.96 (see Note 1).
Note 1: does not include data for 0% side-grooved specimens.

used to characterize the slope of the R-curve. $T_{material}$ for non-side-grooved three-point bend specimens had values of 47.0 and side-grooved specimens had values of 26.0. The value of $T_{material}$ was found to depend upon specimen geometry. The average tearing modulus for compact tension specimens was 17.80 whereas the average value for three-point bend specimens was 26.3. The tearing modulus on average was 47% higher for three-point bend specimens compared with CT specimens. In either specimen geometry the tearing modulus was independent upon starting a/W ratio when side groove depths of 12.5% and 25% were used. Thus it appears that the material tearing modulus should be referred to as $T_{material,geometry}$ since it appears not to be a material property.

Data in Figs. 5–11 and Tables 5 and 6 show that J_{Ic} is a valid fracture parameter since its value is independent of specimen geometry and starting crack length as long as plane-strain conditions are met. Plane-strain conditions are obtained in the three-point bend geometry by the use of side grooves of either 12.5% or 25% depth. In the compact tension specimens the plane-strain conditions are met even when side grooves are not used. The fact that the thickness and width of the CT specimens were twice that of the three-point bend specimen was probably responsible for plane-strain conditions to be maintained in the CT specimens even when side grooves were not used, whereas the three-point bend specimens required side grooves to maintain plane-strain conditions.

Discussion

The accuracy of the crack length measurement obtained by unloading elastic compliance depends upon many variables. Effects of friction in the load train and friction in the knife edge attachment have been examined by Hewitt [12]. Hewitt found that by reducing the pin friction by the use of needle bearings, the crack length error could be reduced from 0.05 to 0.03 mm. The effect of knife edge friction was determined to be small. Four other factors which will effect the crack length accuracy are the error in load measurement, the error in COD measurement, the error in modulus measurement, and the error in the fit of crack length with compliance. Assuming that the polynomial fit of crack length versus compliance is good, one can neglect this error. The error in crack length can be estimated by forming the partial differential of Eq 7. The error in crack length Δa_{error} is given by [31]

$$\Delta a_{\text{error}} = W\left(\frac{\partial a}{\partial U}\right)\sqrt{\left[\left(\frac{\partial U}{\partial E}\,\Delta E_{\text{er}}\right)^2 + \left(\frac{\partial U}{\partial P}\,\Delta P_{\text{er}}\right)^2 + \left(\frac{\partial U}{\partial V}\,\Delta V_{0\text{er}}\right)^2\right]} \quad (10)$$

where U is defined by Eq 8.

Performing the appropriate partial differentiation and defining the variable z as

$$z = \frac{\left(\dfrac{E' V_0 B_{\text{eff}}}{P}\right)^{1/2}}{\left[1 + \left(\dfrac{E' V_0 B_{\text{eff}}}{P}\right)^{1/2}\right]^2} \quad (11)$$

Eq 10 may be reduced to

$$\Delta a_{\text{error}} = \frac{W \cdot z}{2}\,[C_1 + 2C_2 U + 3C_3 U^2 + 4C_4 U^3 + 5C_5 U^4]$$

$$\cdot \left[\left(\frac{\Delta E_{\text{er}}}{E}\right)^2 + \left(\frac{\Delta V_{0\text{er}}}{V_0}\right)^2 + \left(\frac{\Delta P_{\text{er}}}{P}\right)^2\right]^{1/2} \quad (12)$$

In order to evaluate Eq 12 a computer program was written to apply this equation to actual J-resistance test data. The load and COD errors were defined in terms of a percentage error of the test range used. A typical error was on the order of 0.10% of the test range. Since not all of the unloading data were used to calculate crack length, but rather only the section of the load-COD curve that was between the upper and lower limits, the amount that this reduced the load and COD signals was included in the analysis. The influence of load and COD ranging was included in the analysis by the amount that ranging changed the relative load and COD errors.

The results of the error analysis was a tabular output and a graphic output in which error bars were placed on J-Δa_p data pairs. The program was run to determine the effect of load and COD ranging on crack length error. In the three-point bend specimens the use of load and COD ranging reduces the average crack length error from 0.10 to 0.007 mm. In the compact tension specimen the average crack length error is reduced from 0.28 to 0.03 mm by the use of ranging. The analysis has also shown that crack length can be determined more accurately in the three-point bend specimen than in the CT specimen.

An alternative estimate of the error in crack length measurement by unloading compliance was determined using a statistical analysis of the load and COD data. The crack length was determined from the slope of the COD-load curves; thus any error in determining the slope of the curve would result in an error in the crack length. The slope of the COD-load curve was determined by a linear regression analysis of the data. The error in the least-squares slope was determined by the standard error estimate of the slope, S_b [32,33]. The standard error estimate determined the variability in the slope due to the distribution of data points about the least-squares fitted line. The standard error estimate would be large when there is a loose dispersion of data points about the regression line, whereas the standard error estimate would be small when there is a tight dispersion of data points about the regression line. The standard error estimate is defined by the following expressions

$$ S_b = \frac{S_{y/x}}{S_x \sqrt{n-1}} \tag{13} $$

where

$$ S_x = \frac{1}{n(n-1)} [n\Sigma x^2 - (\Sigma x)^2] \tag{14} $$

$$ S_{x/y} = \frac{n-1}{n-2} (S_y^2 - \beta^2 S_x^2) \tag{15} $$

$$ S_y = \frac{1}{n(n-1)} [n\Sigma y^2 - (\Sigma y)^2] \tag{16} $$

The error in crack length caused by the distribution of data about the fitted line can be estimated by

$$\Delta a_{error} = C_0 + C_1 U_{eff} + C_2 U_{eff}^2 + C_3 U_{eff}^3 + C_4 U_{eff}^4 + C_5 U_{eff}^5 \quad (17)$$

The transfer function, U_{eff}, is an effective transfer function which includes the measured slope V_0/P and the estimate of the error in this slope, S_b. The effective transfer function is defined by

$$U_{eff} = \cfrac{1}{\left[E' B_{eff} \left(\cfrac{V_0}{P} + S_b \right) \right]^{1/2} + 1} \quad (18)$$

The error analysis was performed in the real-time test program JRTEST for every unloading compliance measurement. This analysis was referred to as the on-line error analysis, whereas the previous analysis was referred to as the theoretical error analysis.

The errors obtained using both analyses are compared in Table 7. Errors reported are the average error for all the unloading compliance measurements within a given test. The $+/-$ error is the standard deviation for that particular average. A number of interesting comparisons can be made with reference to Table 7. First, consider the effect of load and COD ranging on the magnitude of crack length error. Both error analyses show that the crack length error is reduced substantially by the use of ranging, though the theoretical analysis shows that the difference is far greater than the on-line analysis shows. A comparison can also be made with regard to the effect of specimen type on the crack length error. The on-line analysis showed that the crack length error was similar for the two types of specimens, whereas the theoretical analysis shows that the three-point bend specimen has less error.

The reasons for the differences in the results of the two types of error analysis are due to the differences in how the error was calculated. In the on-line analysis, error was determined by the repeatability and the amount of noise in the measurement. Thus a measurement in which all the load-COD data fall exactly on the fitted regression line would give an error of zero. This analysis

TABLE 7—Comparison of crack length errors using two types of analysis.

Specimen Type	Ranging Used	On-Line Error Analysis Crack Length Error, mm	Theoretical Error Analysis Crack Length Error, mm
Bend	yes	0.0135 ±0.0055	0.00695 ±0.00145
Bend	no	0.0445 ±0.0275	0.108 ±0.035
CT	yes	0.0165 ±0.0055	0.028 ±0.008
CT	no	0.0440 ±0.018	0.2835 ±0.0815

does not consider the absolute accuracy of each measurement but rather only the amount of noise on the measurement. In contrast to this analysis, the theoretical analysis neglects noise on individual data points and concentrates on the accuracy of the individual load and COD signals. The theoretical analysis should give a better estimate of the crack length error, since the source of the crack length errors is used in the analysis.

A graphical representation of the effect of load and COD ranging on the J-R curve is shown in Figs. 12 and 13. In these figures error bars were placed on the physical crack advance data using Eq 12 to calculate the error. Figure 12 shows the results for a test which used ranging. The crack length error is greatest at the beginning of the test when the loads are the smallest. In Fig. 13 the results are plotted for two cases. In the first case, labeled "theory," the test data from a specimen which used ranging were input into Eq 12 and the error was calculated assuming ranging was not used. The curve labeled "experimental" contains test data from a specimen in which ranging was not used in the real-time test. This figure compares the experimentally measured error with the calculated error. The results show that the variation of crack advance seen in experimental data falls within the theoretically calculated error bars. The comparison is clearest along the blunting line.

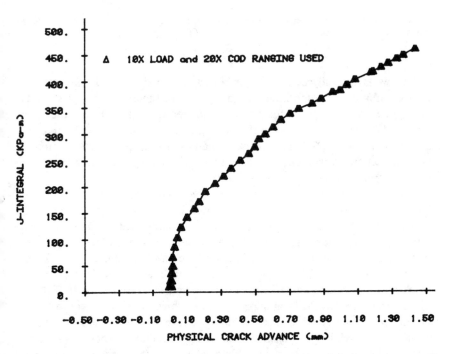

FIG. 12—*J-R curve for 17-4-Ph stainless steel with 0% side grooves using* ×*10 load and* ×*20 COD ranging with error bars calculated by Eq 12.*

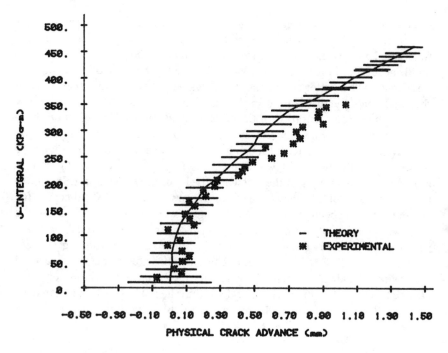

FIG. 13—*Comparison of theoretical and experimental errors on the J-R curve of 17-4-Ph stainless steel with 0% side grooves using no load or COD ranging.*

The two error analyses estimate the crack length error in a J-resistance test by two different but complimentary methods. The theoretical analysis (Eq 12) determines the accuracy of the crack length measurement. The on-line analysis (Eq 17) determines the relative amount of noise in the individual data points of load and COD. The theoretical analysis is useful in determining the absolute accuracy, whereas the on-line analysis is useful in evaluating the effect of system noise on crack length error.

A series of measurements was made on the fractured specimens to compare the optical crack length with that obtained by unloading compliance. The optical crack length was measured at five equidistant locations along the fracture surfaces and then the measurements were averaged. In the compact tension specimens both the initial fatigue precrack and the static J_{Ic} crack were measured, whereas in the three-point bend specimens, only the fatigue precrack was measured. The results of this comparison are shown in Table 8. The average error in the fatigue precrack length was 0.53% for compact tension specimens and 0.83% for three-point bend specimens. The average error in the static crack length was 0.58% for the CT specimen. These results show that the computerized system was able to accurately measure the crack length

TABLE 8—Comparison between crack length measured by compliance and optically.

Specimen No.	Type	% Side Grooves	Precrack Compliance, mm	Precrack Optical, mm	% Difference[a] in Precrack Length	Static Crack Compliance, mm	Static Crack Optical, mm	% Difference[a] in Static Crack Length
5517-4-1	CT	0	31.496	31.293	−0.65	34.899	34.392	−1.45
5517-4-2	CT	0	35.585	35.507	−0.22	37.338	...	...
5517-4-3	CT	12.5	35.789	35.992	+0.57	39.421	39.980	+1.42
5517-4-4	CT	12.5	40.767	40.864	+0.24	41.808	41.681	−0.30
5517-4-5	CT	12.5	35.687	35.743	+0.16	39.116	39.192	+0.19
5517-4-6	CT	12.5	38.024	38.231	+0.53	41.250	41.438	+0.46
5517-4-7	CT	25	35.484	36.052	+1.60	39.929	40.192	+0.66
5517-4-8	CT	25	40.630	40.731	+0.25	43.307	43.373	+0.15
5517-4-9	CT	25	38.049	38.173	+0.33	40.513	41.092	+1.50
5517-4-16	bend	0	17.907	17.478	−2.40	20.269	...	...
5517-4-17	bend	0	20.422	19.921	−2.45	21.793	...	...
5517-4-19	bend	25	20.268	20.688	+2.08	22.556	...	...
5517-4-20	bend	25	17.831	18.019	+1.05	22.098	...	...
5517-4-21	bend	25	19.126	19.070	−0.29	22.428	...	...
5517-4-22	bend	12.5	17.729	18.059	+1.86	22.047	...	...
5517-4-23	bend	12.5	19.177	19.456	+1.46	22.403	...	...
5517-4-24	bend	12.5	16.586	16.805	+1.32	21.590	...	...
5517-4-25	bend	12.5	20.371	20.417	+0.22	21.793	...	...

[a] % Difference $= \dfrac{\text{(optical crack length} - \text{compliance crack length)}}{\text{compliance crack length}} \times 100$

CT specimen average precrack length error = +0.53%.
CT specimen average static crack length error = +0.58%.
Three-point bend specimen average precrack length error = 0.86%.

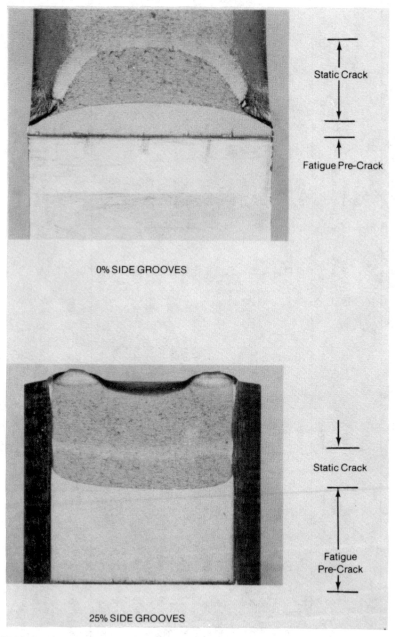

FIG. 14—*Fracture surface of compact tension specimens with no side grooves and with 25% side grooves.*

and that the use of the effective thickness in Eq 8 does account accurately for the effect of side grooves on the unloading elastic compliance. The five-point optically measured crack length was always less than that measured by compliance with non-side-grooved specimens. This is primarily due to the significant amount of tunneling observed in non-side-grooved specimens. The crack front profiles are shown in Fig. 14 for the compact tension specimens. In the CT specimens with 0% side grooves there was some tunneling observed in the fatigue precrack front, but there was a significant tunneling observed in the static crack front. In the case of 25% side grooves the static crack front was straight and the fatigue precrack had an inverse curvature with the crack length being greater at the surface.

The static crack front did not maintain a planar front when side grooves were not used; the surface regions of the crack propagated by a shear mode as evidenced by the large shear lips and the center region propagated in a planar Mode I plane-strain fracture path. This mixed-mode crack advance was the result of a change in the stress state from plane stress at the surface to plane strain in the center. It was estimated that 35% of the crack was in plane stress and 65% was in plane strain. In side-grooved specimens, 100% of the crack front was in plane strain.

Conclusions

1. Procedures to automate the measurement of J-resistance curves were presented. The automated test controls a screw-driven electromechanical test machine and determines a specimen's J-resistance curve. Compact tension or three-point bend specimens could be tested with the automated system.

2. The accuracy of the crack length measurements made by unloading compliance was determined using two different types of error analysis. Both error analyses showed that load and COD ranging was effective in reducing the crack length errors.

3. The test results showed that J_{Ic} is a valid fracture criterion when plane-strain conditions are met. J_{Ic} was shown to be independent upon starting a/W ratio and specimen type.

4. The J-resistance curves for specimens with side grooves were shown to be unaffected by the side groove depth.

5. The material tearing modulus was not a material property. The tearing modulus was always higher for the three-point bend geometry. It was suggested that the tearing modulus be referred to as $T_{\text{material,geometry}}$.

References

[1] Eshelby, J. D., *Proceedings of the Royal Society of London*, Series A, Vol. 241, 1957.
[2] Hutchinson, J. W., *Journal of the Mechanics and Physics of Solids*, Vol. 16, 1968, pp. 13–31.

[3] Rice, J. R. and Rosengren, G. F., *Journal of the Mechanics and Physics of Solids*, Vol. 16, 1968, pp. 1–12.

[4] Rice, J. R., *Journal of Applied Mechanics*, June 1968, pp. 379–386.

[5] *1981 Annual Book of ASTM Standards*, Part 10, American Society for Testing and Materials, Philadelphia, 1981.

[6] Clarke, G. A., Andrews, W. R., Begley, J. A., Donald, J. K., Embley, G. T., Landes, J. D., McCabe, D. E., and Underwood, J. H., "A Procedure for the Determination of Ductile Fracture Toughness Values Using J-Integral Techniques," *Journal of Testing and Evaluation*, Vol. 7, No. 1, Jan. 1979, pp. 49–56.

[7] Clarke, G. A., "Evaluation of the J_{Ic} Testing Procedure by Round Robin Test on A533B Class 1 Pressure Vessel Steel," *Journal of Testing and Evaluation*, Vol. 8, No. 5, Sept. 1980, pp. 213–220.

[8] Clarke, G. A., Landes, J. D., and Begley, J. A., "Results of An ASTM Cooperative Test Program on the J_{Ic} Determination of HY130 Steel," *Journal of Testing and Evaluation*, Vol. 8, No. 5, Sept. 1980, pp. 221–232.

[9] Clarke, G. A. in *Fracture Mechanics: Thirteenth Conference, ASTM STP 743*, American Society for Testing and Materials, Philadelphia, 1981, pp. 553–575.

[10] Hutchinson, J. W. and Paris, P. C. in *Elastic-Plastic Fracture, ASTM STP 668*, American Society for Testing and Materials, Philadelphia, 1979, pp. 37–64.

[11] Jones, R. L., Duggan, T. V., Spence, L. J., and Barnes, P. J., "Comparison of R-Resistance Curves," *Proceedings*, 4th ECF Conference, Cannes, France, 1982.

[12] Hewitt, R., "Accuracy and Precision of Crack Length Measurement Using Compliance Technique," *Journal of Testing and Evaluation*, Vol. 11, No. 2, March, 1983, pp. 150–155.

[13] Joyce, J. A. and Geidas, J. P. in *Elastic-Plastic Fracture, ASTM STP 668*, American Society for Testing and Materials, Philadelphia, 1979, pp. 451–468.

[14] Van Der Sluys, W. A. and Futato, R. J., "Computer-Controlled Single-Specimen J-Test," *Elastic-Plastic Fracture: Second Symposium, ASTM STP 803*, American Society for Testing and Materials, Philadelphia, 1983, pp. II-464-II-482.

[15] Matthews, J. R. and West, G. D., "J_{Ic} Measurement Point Determination for HY130, CMS-9 and Inconel Alloy 718," *Journal of Testing and Evaluation*, Vol. 11, No. 3, May 1983, pp. 217–224.

[16] Takahashi, H., Khan, M. A., and Suzuki, M., "A Single Specimen Determination of J_{Ic} for Different Alloy Steels," *Journal of Testing and Evaluation*, Vol. 8, No. 2, March 1980, pp. 63–67.

[17] Andrews, W. R. and Shih, C. F. in *Elastic-Plastic Fracture, ASTM STP 668*, American Society for Testing and Materials, Philadelphia, 1979, pp. 426–450.

[18] Underwood, J. H. in *Fracture Mechanics: Eleventh Conference, ASTM STP 677*, American Society for Testing and Materials, Philadelphia, 1979, pp. 463–473.

[19] Milne, J. and Chell, G. G. in *Elastic-Plastic Fracture, ASTM STP 668*, American Society for Testing and Materials, Philadelphia, 1979, pp. 358–377.

[20] Vassalaros, M. G., Joyce, J. A., and Geidas, J. P., in *Fracture Mechanics: Twelfth Conference, ASTM STP 700*, American Society for Testing and Materials, Philadelphia, 1980, pp. 251–270.

[21] Clarke, G. A. and Landes, J. D., in "Toughness Characterization and Specification for HSLA and Structural Steels," *Transactions*, American Institute of Mining, Metallurgical and Petroleum Engineers, 1979, pp. 79–111.

[22] Paris, P. C., Tada, H., Zahour, Z., and Ernst, H. in *Elastic-Plastic Fracture, ASTM STP 668*, American Society for Testing and Materials, Philadelphia, 1979, pp. 5–36.

[23] Paris, P. C., Tada, H., Zahour, A., and Ernst, H. in *Elastic-Plastic Fracture, ASTM STP 668*, American Society for Testing and Materials, Philadelphia, 1979, pp. 251–265.

[24] Ernst, H. A., Paris, P. C., and Landes, J. D. in *Fracture Mechanics: Thirteenth Conference, ASTM STP 743*, American Society for Testing and Materials, Philadelphia, 1981, pp. 476–502.

[25] Joyce, J. A. and Vassdaros, M. G. in *Fracture Mechanics: Thirteenth Conference, ASTM STP 743*, American Society for Testing and Materials, Philadelphia, 1981, pp. 525–542.

[26] Jablonski, D. A., "Automated Crack Growth Rate Testing Using a Computerized Test System," *Proceedings*, SEECO '83, International Conference on Digital Techniques in Fatigue, The City University, London, 28–30 March 1983, pp. 291–308.

[27] Merkle, J. G., "A J-Integral Analysis for the Compact Specimen, Considering Axial Force as well as Bending Effects," *Transactions*, American Institute of Mining, Metallurgical and Petroleum Engineers, Nov. 1974, pp. 286–292.

[28] Saxena, A. and Hudak, S., *International Journal of Fracture*, Vol. 14, 1978, pp. 453–468.

[29] Jablonski, D., "Compliance Method to Determine Crack Length in an ASTM 3-Point Bend Specimen," Instron Corporate Research Lab Report No. G00124, Canton, MA, May, 1982.

[30] Shih, C. F. and deLorenzi, H. G., "Elastic Compliances and Stress Intensity Factors for Size-Grooved Compact Specimens," *International Journal of Fracture*, Vol. 13, pp. 544–548.

[31] Young, H. D., *Statistical Treatment of Experimental Data*, McGraw-Hill, New York, 1962, pp. 3–9 and pp. 96–101.

[32] Dixon, W. J. and Massey, F. J., *Introduction to Statistical Analysis*, McGraw-Hill, New York, 1957, Chapter 11, pp. 189–196.

[33] Crow, E. L., Davis, F. A., and Maxfield, M. W., *Statistics Manual*, Dover, New York, 1955, Chapter 6, pp. 147–194.

Author Index

Subject Index